Kedes Paulo Pereira
Jucelane Salvino de Lima
Dulciene Karla de Andrade Silva

Mineral requirements for goats

Kedes Paulo Pereira
Jucelane Salvino de Lima
Dulciene Karla de Andrade Silva

Mineral requirements for goats

ScienciaScripts

Imprint

Any brand names and product names mentioned in this book are subject to trademark, brand or patent protection and are trademarks or registered trademarks of their respective holders. The use of brand names, product names, common names, trade names, product descriptions etc. even without a particular marking in this work is in no way to be construed to mean that such names may be regarded as unrestricted in respect of trademark and brand protection legislation and could thus be used by anyone.

Cover image: www.ingimage.com

This book is a translation from the original published under ISBN 978-613-9-61705-0.

Publisher:
Sciencia Scripts
is a trademark of
Dodo Books Indian Ocean Ltd. and OmniScriptum S.R.L publishing group

120 High Road, East Finchley, London, N2 9ED, United Kingdom
Str. Armeneasca 28/1, office 1, Chisinau MD-2012, Republic of Moldova, Europe
Printed at: see last page
ISBN: 978-620-7-89747-6

Kedes Paulo Pereira
Jucelane Salvino de Lima
Dulciene Karla de Andrade Silva

Mineral requirements for goats

ScienciaScripts

Imprint

Any brand names and product names mentioned in this book are subject to trademark, brand or patent protection and are trademarks or registered trademarks of their respective holders. The use of brand names, product names, common names, trade names, product descriptions etc. even without a particular marking in this work is in no way to be construed to mean that such names may be regarded as unrestricted in respect of trademark and brand protection legislation and could thus be used by anyone.

Cover image: www.ingimage.com

This book is a translation from the original published under ISBN 978-613-9-61705-0.

Publisher:
Sciencia Scripts
is a trademark of
Dodo Books Indian Ocean Ltd. and OmniScriptum S.R.L publishing group

120 High Road, East Finchley, London, N2 9ED, United Kingdom
Str. Armeneasca 28/1, office 1, Chisinau MD-2012, Republic of Moldova, Europe
Printed at: see last page
ISBN: 978-620-7-89747-6

Table of contents:

Introduction

The Oxford Dictionary defines minerals as a homogeneous solid inorganic substance with a defined physical composition and crystalline structure necessary for the body. Another concept used to define them is that they are the components of animal and plant tissues that make up ash, i.e. the residue that remains after organic matter has been completely burned.

Some time ago, there was a consensus that the nutritional requirements of animals were based mainly on protein and energy requirements. However, with the introduction of new technologies, research has improved, providing more accurate and precise answers to the nutrient requirements of animals, and it has been possible to verify that in addition to protein and energy, other compounds, such as minerals, are also necessary to meet the requirements for maintenance and production, such as: organ function, tissue maintenance and the production of meat, milk and milk.

It's a fact that minerals are an important component of goats' diets and have a strong influence on production rates. Knowledge of mineral requirements is essential in order to develop supplements that can adequately meet the nutritional needs of these animals.

However, mineral requirements are highly dependent on the level of productivity. Higher rates of weight gain or higher milk production also require higher quantities of minerals. Therefore, management practices that lead to higher milk production and animal growth rates imply greater attention to meeting mineral requirements.

However, the variation in the quantities required for many minerals is affected by both the diet and the animal. But, especially in grazing regimes, when animals do not receive concentrated feed or receive it in limited quantities, they depend on the concentrations of minerals present in grasses, which, in turn, present limited quantities of many of these elements. However, supplementation with macro and microminerals is extremely important for the production process (Pedreira & Berchielli, 2006).

However, the use of technologies such as pasture supplementation in a semi-confinement system is carried out by estimating pasture consumption, in order to know how much nutrients the animals are taking from the pasture, and then the supplement is provided to make up the deficit and meet their requirements. However, for animals in confinement, the diet provided will be the only source of nutrients, which must contain all the minerals required by the animal in order to achieve maximum performance.

However, some factors can influence the supply of these nutrients, leading to a nutritional imbalance and having a marked influence on productivity. Factors such as soil fertility, plant species and climatic conditions must be taken into account (Costa et al., 2004). Estimating the amount of minerals available in the environment is essential if mineral supplements are to be formulated efficiently. This attention is reported by Teixeira (1992), when he emphasizes the fundamental importance of minerals in animal nutrition, as they are required for various biological processes, such as the structural composition of the body, participation in the ionic form of intra- and extracellular fluids and liquids, as an enzyme catalyst and in hormonal functions.

Several strategies have been adopted to better meet the mineral requirements of goats, which, if not met, can lead to various metabolic alterations, directly interfering with productive performance (Pedreira and Berchielli, 2006).

Recently, studies into bioavailability have shown that the metabolism of minerals cannot be considered in isolation. Physiological and nutritional factors can interfere with absorption, transport and storage, with subsequent changes in the requirements of these elements.

The bioavailability of minerals in food, used in the basal diet and in supplementation, is considered an essential factor for good mineralization. Bioavailability means the amount of a nutrient that is available for absorption in the form in which it is physiologically usable (Filisetti and Lobo, 2007).

For McDowell (1999), it was only when methodologies were created to study minerals in animal tissue and forage, observing the animals' responses to the supplementation of isolated minerals, that it was possible to replace assumptions with facts about the essentiality of mineral nutrients.

However, the use of data on mineral requirements for goats is still usually obtained from tables drawn up outside Brazil, due to the fact that the data obtained from research carried out in the country is still small in volume, which is a limiting factor in drawing up a table of requirements under Brazilian conditions. For this reason, the results obtained in experiments with increasing levels are still of great value, as they are an important reference tool for correct mineralization, although the recommended supplementation may need to be adjusted.

Currently, the focus of research aimed at determining the mineral requirements of animals is not only on meeting goats' needs for these elements, avoiding imbalance (deficiency or excess), but also on reducing the levels of these elements in the diet, with the aim of reducing production costs and also the excretion of inorganic elements into the environment without, however, affecting animal performance.

Chapter 1

Availability of minerals

In Brazil, the most sustainable form of goat production is based on pasture, because pasture is considered a low-cost source of food. However, the concentration of minerals in this fodder depends on various factors linked to the soil, plants and climate.

Corroborating this, Gomide (1976) states that the mineral composition of forage plants can vary depending on a series of interdependent factors, including: the age of the plant, the soil and the fertilizations made, differences between species and varieties, seasons and succession of cuts.

Low concentrations of mineral nutrients in fodder may be due to the low availability of the mineral in the soil, or may also be due to the plant's reduced genetic capacity to accumulate the element, or may be indicative of the low requirement of the mineral element for plant growth. In the same way, high concentrations or toxic levels of some minerals in the composition of forage are indicative of excess availability in the soil, the plant's genetic or physiological capacity to accumulate high rates of these elements, or high requirements for growth (Underwood, 1983).

In relation to the soil, for example, changes in pH can alter both the concentration of minerals and their availability, as happens in acidic soils, where it can influence the absorption of Ca, Mg and P by plants, which influences the supply of minerals in pastures, their availability to animals and consequently nutritional requirements (Amado, 2002).

According to Van Soest (1994), the effects of soil on forage can be assessed from the point of view of the accumulation of minerals in plants, and the influence of minerals on the yield, composition and digestibility of organic matter in forage, where plants growing on different soils have different mineral balances, altering their composition and growth.

However, the distribution of minerals in plants can also vary in terms of bioavailability, as it is related to the chemical structural form in the different anatomical parts (Whitehead et al., 1985).

Table 1 shows the essentiality of some minerals for animals and plants.

Table 1. Essential elements in animal nutrition and forage plants

Element _	Essentiality		Element _	Essentiality	
	Animals	Plants		Animals	Plants
Ca	+		Cu	+	+
P	+	+	Mn	+	+
K	+	+	Co	+	
In	+	+	Zn	+	+

Cl	+	+	If	+	
Mg	+	+	Mo	+	+
S	+	+	B		+
I	+		Cr	+	
Fe	+	+	F		+

Adapted from Metry (2003).

In general, plants require high proportions of potassium, calcium, phosphorus, magnesium and sulphur (macro-elements), and small amounts of iron, copper, manganese, molybdenum, zinc, chlorine and boron (micro-elements). However, plants and animals differ in their requirements for specific minerals. Animals are not very demanding of boron, but they do need high amounts of sodium and chlorine, as well as small proportions of cobalt, selenium, iodine, nickel and chromium, in addition to those minerals required by plants (Heath et al., 1985).

The absorption of minerals by plants depends on the elements in the soil, their availability and the plant's ability to absorb them. However, the availability of minerals depends on numerous factors such as temperature, humidity and organic matter content. Intrinsic plant factors such as genetic potential and physiological conditions also have a marked influence on the absorption and inorganization of minerals.

The maturation process, which is accompanied by a reduction in nutritional value, can be accelerated by light, temperature and humidity, while it can be slowed down by cutting or grazing. Leaves, sheaths and stems are similarly digestible during the initial growth phase. However, stems mature earlier and their components change with age. As the plant matures, there is a higher concentration of lignin, which binds to most minerals, making them unavailable to the animal due to the difficulty in digesting the fiber by rumen microorganisms. This is confirmed by Powell et al. (1978) who observed a tendency for the use of mineral elements by the animal to fall as the plant matures in temperate forages and tropical grasses.

Table 2 shows the average mineral content of tropical and temperate forages.

Table 2. Average content of some minerals in the dry matter of grasses and legumes from temperate and tropical climates.

	Temperate climate		Cliimi Tropical	
	Grass	Leguminous	Grass	Leguminous
Phosphorus (%)	0.33 (400)	0.36 (320)	0.22 (586)	0.26 (165)

Calcium (%)	0.59 (428)	1.86 (291)	0.40 (390)	1.21 (154)
Magnesium (%)	0.18 (335)	0.29 (193)	0.36 (280)	0.40 (48)
Sodium (%)	0.23 (318)	0.19 (121)	0.26 (192)	0.07 (40)
Copper (ppm)	6 (127)	12 (93)	15 (94)	10 (17)
Zinc (ppm)	32 (31)	55 (34)	36 (119)	42 (7)
Cobalt (ppm)	0.20 (111).	0.42 (21)	0.16 (45)	0.07 (3)

Norton (n.d.), quoted by Dias (1997). Values in brackets = no. of observations. Butler and Junes (1973) and Ha/ell (1985), cited by Véras (1999).

When evaluating the vegetation and climate conditions of Brazil's Northeast region, for example, it can be said that it is a region marked by irregular rainfall with low rainfall rates, which hampers the development of pastures, especially in the semi-arid regions where 90% of Brazil's goat herds live. However, in these conditions, herds tend to have a limited supply of minerals (Table 3), due to the limitations imposed by the environment, making it difficult for the animals to meet their mineral requirements.

Table 3. Mineral composition of some cacti in the semi-arid region of Paraiba.

Cactaceae	Photophore	Potassium	Calcium	Magnesium
Giant palm	0,13	2,89	1,82	0,67
Sweet palm	0,20	3,53	3,84	0,87
Mandacarú	0,06	1,67	2,80	0,41
Knifemaker	0,12	2,13	5,03	1,43
Xiquexique	0,08	1,85	3,74	2,14
Friar's crown	0,17	3,95	2,06	1,04

However, the biochemical bioavailability of minerals is directly linked to the physiological functions in which they participate, resulting in large variations in the absorption coefficients used. The NRC (2007) itself uses absorption coefficients from the AFRC (1991) and others from the NRC (1985) in certain equations for sheep. However, for the equations for goats, generalized data is still used for absorption coefficients and the same for different physiological situations of the animals (maintenance, gain, lactation and pregnancy) due to the lack of sufficient data for this species, often having to resort to data for sheep and cattle suggested by the different committees.

However, another factor that can influence bioavailability is during rumen fermentation. One example is the phytase produced by rumen bacteria. In the rumen, phytase hydrolyzes phytate to phosphoric acid and inositol. When phytate is broken down, the resulting P becomes available to the animal (Lofgreen & Kleiber, 1953; Morse et al., 1992).

Table 4 shows the concentration of minerals in some forage plants during the two seasons.

Table 4. Average concentration of mineral elements in different forage species

Forage species	Ca (%)		P (%)		Mg (%)		S (%)		K (%)	
	Water	Drought	Water	Drought	Water	Drought	Water	Drought	Water	Drought
B. BRIZANTHA	0,29	0,40	0,13	0,11	0,29	0,37	0,14	0,12	1,86	1,16
B. DECUMBENS	0,26	0,33	0,13	0,09	0,26	0,26	0,13	0,12	1,74	1,15
B. HUMIDICOLA	0,30	0,26	0,14	0,11	0,20	0,25	n.a.	n.a.	0,76	0,30
Colonization	0,26	0,46	0,17	0,12	0,22	0,29	0,17	0,16	1,74	1,30
Tobiata	0,27	0,40	0,14	0,10	0,19	0,22	0,15	1,13	1,68	1,29
Tanzania	0,30	0,43	0,15	0,11	0,24	0,28	0,15	0,13	1,66	1,22
Species	Fe (mg/kg)		Mn (mg/kg)		Zn (mg/kg)		Cu (mg/kg)		Na (mg/kg)	

forage	Water	Drought	Water	Drought	Water	Drought	Water	Drought	Water	Dry
B. BRIZANTHA	406	n.a.	107	n.a.	20	25	6,0	6,0	58	43
B. DECUMBENS	223	251	189	201	21	20	5,4	5,2	97	89
B. HUMIDICOLA	441	160	265	222	27	29	3,8	2,7	2465	1214
Cologne	281	446	102	203	22	24	8,9	8,5	99	96
Tobiata	960	1929	142	157	18	17	9,7	8,7	46	57
Tanzania	200	570	145	267	16	15	6,8	6,8	79	94

Adapted from Morais (2001).

Table 5 shows the percentage of some minerals and sources used as supplements and their respective bioavailability.

The NRC (1996), reporting on the importance of mineral elements for ruminants, emphasizes that microbial activity is closely linked to the intake of minerals by the animals, where under conditions of deficiency, what can be observed is a drop in microbial growth and a reduction in food digestibility, depending on the severity of the mineral deficiency (Leng, 1990; Spears, 1994) and the availability of the mineral.

Table 5. Percentage of minerals in some sources used in supplements and their relative availability

Element	Source	% element in source	Bioavailability
Ca	Autoclaved bone meal	29 (23-37)	High
	Defluorinated rock phosphate	29,2 (19,9-35,7)	Intermediate

	Calcium carbonate	40	Intermediate
	Soft phosphate	18,0	Low
	Calcitic limestone	38,5	Intermediate
	Dolomitic limestone	22,3	Intermediate
	Monocalcium phosphate	16,2	High
	Tricalcium phosphate	31,0-34,0	
	Bi-calcium phosphate	23,2	High
	Hay in general		Low
	Calcium sulphate	20,0	Low
P	Defluorinated rock phosphate	13,1 (8,7 - 21,0)	Intermediate
	Calcium phosphate	18,6 - 21,0	High

	Bicalcium phosphate	18,5	High
	Tricalcium phosphate	18,0	
	Phosphoric acid	23,0 - 25,0	High
	Sodium phosphate	21,0 - 25,0	High
	Potassium phosphate	22,8	
	Soft phosphate	9,0	Low
	Autoclaved bone meal	12,6 (8-18)	Intermediate
Mg	Magnesium carbonate	21,0 - 28,0	High
	Magnesium chloride	12,0	High
	Magnesium oxide	54,0 - 60,0	High
	Magnesium sulphate	9,8 - 17,0	High

S	Calcium sulphate (gypsum)	12,0 - 21,0	Low
	Potassium sulphate	28,0	High
	Magnesium potassium sulphate	22,0	High
	Sodium sulphate	10,0	Intermediate
	Anhydrous sodium sulphate	22,0	
	Sulphur flower	96,0	Low
K	Potassium chloride	50,0	High
	Potassium sulphate	41,0	High
	Potassium sulphate	18,0	High
Co	Cobalt carbonate	46,0 - 55,0	High
	Potassium sulphate	41,0	High

	Potassium sulphate	18,0	High
Cu	Copper sulphate	25,0	High
	Copper carbonate	53,0	Intermediate
	Copper chloride	37,2	Intermediate
	Copper oxide	80,0	Low
	Copper nitrate	33,9	Intermediate
Fe	Iron oxide	46,0 - 60,0	Not available
	Iron carbonate	36,0 - 42,0	Low
	Iron sulphate	20,0 - 30,0	High
I	calcium Iodate	63,5	High
	Stabilized potassium iodate	69,0	High

	Copper iodide	66,6	High
	Ethylenediamine	80,0	High
	dihydroiodide		
Mn	Manganese sulphate Manganese oxide	27,0 52,0 - 62,0	High Intermediate
If	Sodium selenate	40,0	High
	Sodium selenite	45,6	High

Adapted from McDowell (1999).

Table 6 shows some of the metabolic functions of minerals, strengthening the idea of the importance of minerals for ruminants.

Table 6. Some of the most important metabolic functions of some minerals

Mineral	Composition in the body (%)	Fungi
Calcium (Ca)	1 - 2	Bone formation, metabolic regulation, blood coagulation, muscle contraction, nerve impulse transmission
Phosphorus (P)	0,7 - 1,2	Bone formation, component of DNA and RNA, ATP, regulation of allosteric enzymes, phospholipids

Potassium (K)	0,3	Osmotic pressure, transmission of nerve impulses, regulation of acid-base balance, muscle contraction, control of water balance
Sulphur (S)	0,25	Sulfur aa component, vitamin and mucopolysaccharide component, detoxification reactions
Sodium (Na)	0,15	Regulation of osmotic pressure, conduction of nerve impulses, active transport, regulation of acid-base balance, muscle contraction, control of water balance
Chlorine (Cl)	0,15	Regulation of osmotic pressure, regulation of acid-base balance, control of water balance, formation of HCl in gastric juice
Magnesium (Mg)	0,045	Cofactor of more than 300 enzymes, component of bones and neuro-muscular activities

Adapted from SPEARS (1999).

Data from the literature has indicated mechanisms of interaction between minerals and microorganisms in the gastrointestinal tract. However, mineral elements are essential, both for microorganisms and for the animal organism, exerting, for example, a direct effect on meat quality by promoting tissue growth in a way that determines the qualitative aspects of animal carcasses (Pedreira & Berchielli, 2006).

Estimating mineral requirements

The vast majority of specialized literature on the issues surrounding nutritional requirements always highlights the factors that can influence the mineral requirements of animals, such as: production levels, the structure of the elements present in forage (bioavailability), interrelationships with other nutrients and between minerals, breed, sex, age, body composition, physiological state and other factors such as: temperature, dryness, relative humidity, wind speed and luminosity (Conrad et al., 1985; Maynard et al., 1984; Sousa et al. 1998).

However, this range of variables forms a barrier, making it difficult to effectively understand the mineral requirements for animals in Brazilian conditions, which is perhaps one of the causes for the low number of studies on mineral requirements for goats, preventing the formation of tables that are characterized and based on Brazilian conditions, leading to the use of North American and European systems to formulate diets for goats in the country.

The fact is that recommendations from foreign committees are currently being used, although there are doubts about their effectiveness, increasing the risk of overestimating or underestimating mineral requirements for goats raised in tropical regions.

It is also true, as previously reported, that environmental conditions and animal characteristics affect mineral requirements. However, in view of the scarcity of representative data on goats, the committees themselves sometimes resort to using values derived from requirements for cattle and sheep, further increasing the likelihood of errors.

However, currently, despite the difficulties, several experiments have been carried out in Brazil to estimate the mineral requirements for goats, such as those carried out by Sousa et al. (1998); Bueno and Vitti (1999); Geraseev et al. (2000); Queiroz et al. (2000); Carvalho et al. (2003); Costa et al. (2003), in order to serve as a data base for formulating a table of their own in the future.

The first recommendations on minerals came from experiments involving the relationship between mineral concentrations in feed and animal performance. The requirements were defined by the need for maximum performance and maximum efficiency of feed utilization. However, these experiments demonstrate mineral requirements in an empirical way, because if there are any changes in either the feed or the animals, these evaluations become inconsistent (Costa, 2006).

Various methods have been proposed for estimating mineral requirements, one of which is the factorial method. To use this method, endogenous losses are first calculated, the purpose of which is to estimate net mineral requirements for maintenance through the amount of minerals excreted by the gastrointestinal tract that has not been reabsorbed. The amount used by the animal for growth, fattening, pregnancy, milk production and milk growth is then calculated. The sum of the net requirements for maintenance and production will constitute the total net requirement, followed by a correction of the absorption coefficient of each element in the animal's gastrointestinal tract, resulting in the mineral's dietary requirement (Silva, 1995).

Another method used is comparative slaughter, which is currently the most widely used and has produced results that form the basis of the Californian System (Lofgreen & Garrett, 1968). However, although the comparative slaughter method is effective, it is somewhat expensive.

The methodology of using radioisotopes can also be used, where mineral requirements are obtained through the accumulation of these stable radioactive isotopes in certain organs, where intrinsic and extrinsic isotopes can be used for this purpose. Intrinsic isotopes can be present in plants or in the animal itself, resulting in the marking of the tissue in which they have been retained. However, it is assumed that the isotope and the mineral are used in the same way by the animals. In the case of extrinsic isotopes, they must be introduced into the food and the amount of mineral supplied must also be consumed, resulting in the marking of animal tissue. These radioactive elements are used to quantify the endogenous fraction (Comar, 1953). The major limitation of this method is the cost of carrying it out and the handling of radioactive materials.

The NRC (2007), unlike the other NRCs, calculates all the mineral requirements, considering 14 essential minerals for goats and sheep, namely: calcium, phosphorus, chlorine, sodium, potassium, magnesium, sulphur, copper, cobalt, iron, manganese, selenium, iodine and zinc, not including molybdenum and fluoride as microminerals, which, in the NRC (1981), are considered essential mineral elements. The NRC (1985), on the other hand, cites 15 mineral elements, as it also includes molybdenum as a micromineral.

In addition to these, there are indications of the essentiality of other minerals such as: aluminum (Al), arsenic (As), boron (Bo), chromium (Cr), fluorine (F), lead (Pb), lithium (Li), molybdenum (Mo), nickel (Ni), rubidium (Rb), silica (Si), tin (Sn) and vanadium (V); However, the study of these elements does not seem to be very intense, as their dietary requirements are extremely low with little possibility of deficient levels occurring in normal diets, and they are better known for their toxic effects: fluoride (F), molybdenum (Mo), lead (Pb), arsenic (As), aluminum (Al), cadmium (Cd) and mercury (Hg) (NRC, 2007).

In addition to these differences with the other committees and the previous NRCs, the NRC (2007) more clearly addresses the metabolism and functions of all minerals, as well as the problems caused by excess or deficiency, which are not presented with much emphasis in the others.

Calcium (Ca) requirements

Calcium is considered a fundamental mineral for the animal organism, being directly linked to bone composition, from its formation, as well as its maintenance throughout life with the removal and replacement of this element in bone tissue. According to Capen (1980), calcium is involved in regulating various organic processes, such as blood coagulation, transport through membranes, secretory processes, enzymatic reactions, neuromuscular excitability, the release of hormones and neurotransmitters, as well as the intracellular action of various hormones.

Calcium is a bivalent cation that is abundant in the animal body and alone accounts for 1 to 2% of body weight. Around 99% of total calcium is found in bones and teeth. In bone, calcium is in the form of hydroxyapatite (Ca10(PO4)6(OH)2), a crystalline structure consisting of calcium phosphate arranged around an organic matrix of collagenous protein to provide strength and rigidity. This same type of crystal is present in tooth enamel and dentin (Dell'Isola et al., 2003).

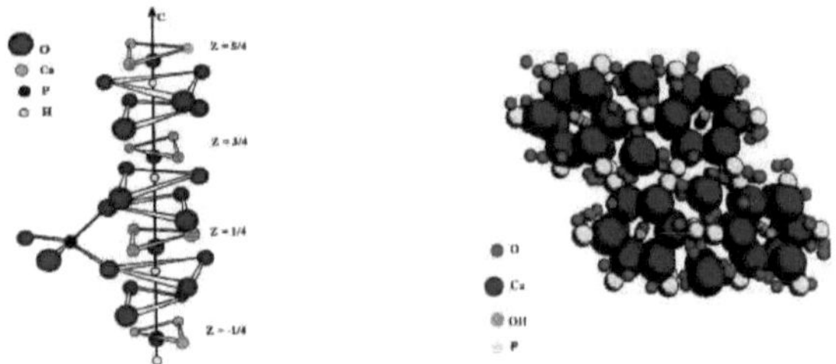

Figure 1 - Structure of Hydroxyapatite Fiocruz (1999).

García et al. (2002) reports that 1% of Ca is present in other tissues and organs and is responsible for various metabolic processes such as blood clotting and muscle excitability (Holick, 1999).

Calcitonin, calcitriol (the active form of vitamin D3) and parathormone (PTH) are the main hormones involved in the Ca balance. When serum Ca concentrations decrease in the blood, PTH converts vitamin D3 into calcitriol in the kidneys (Figure 2).

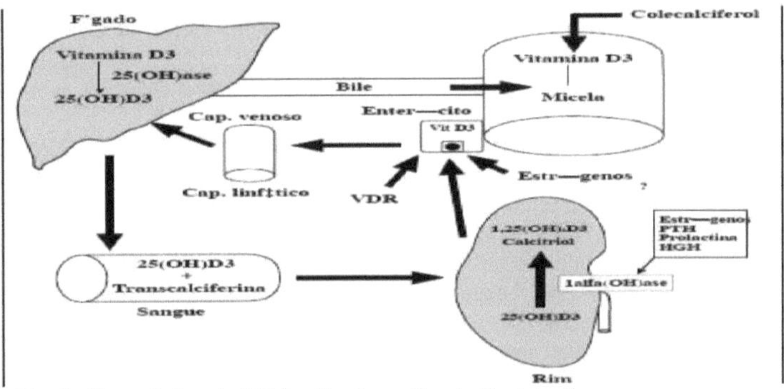

Figure 2 - Metabolism of vitamin D3 for Ca absorption in the intestine.
http://www.puravitamina.com.br

Calcitriol increases Ca absorption in the small intestine (the site of greatest Ca absorption), as this hormone stimulates intestinal cells to synthesize two Ca transport proteins, Calbidin (CaBP), which transports Ca into the intestinal lumen. Calcitriol and PTH also stimulate osteoclast activity in the bones and increased Ca absorption in the kidneys. Calcitonin is the antagonistic hormone and prevents the physiological development of hypercalcemia by inhibiting osteoclastic and renal Ca reabsorption. According to the NRC (2007), body Ca is excreted mainly through feces (85-98%) and urine (2-5%), in addition to sweating.

The calcium absorption mechanism is complex, involving various factors such as: vitamin D, ATPase, intestinal alkaline phosphatase, factors that increase or decrease its solubility, calcium-binding protein in the enterocyte (clabindin), calcium-binding protein in the plasma and others.

Ca absorption is dependent on acidic pH, which is why it occurs more intensely in the duodenum and decreases in the rest of the intestine as alkalinity increases (Bronner, 1998). Heaney et al. (2000) reports that in an alkaline environment there is a breakdown of the calcium-phosphate bond, forming calcium phosphate, this element is considered insoluble and is subsequently eliminated through the feces.

The intestinal absorption of Ca can be divided into two parts: an active saturable part, which is mediated by vitamin D and involves the Ca-binding protein (Capen, 1980), and a passive part, which can correspond to simple or facilitated diffusion.

The body preserves plasma calcium, so around 90% is reabsorbed in the proximal tubules, the henle's flaps and the initial portion of the distal tubules. The reabsorption of the remaining 10% is very selective, depending on the concentration of calcium ion in the blood. When it is low, this reabsorption is great, so that almost no calcium is lost in the urine. On the other hand, a small increase in the concentration of calcium ion in the blood above normal sharply increases calcium excretion (Guyton, 1997).

Vitamin D is a steroid hormone that exists in the skin and is activated by ultraviolet rays, where it undergoes heat-induced isomerization over a few hours and is transformed into vitamin D3. On reaching the bloodstream, vitamin D3 is immediately transported to the liver to undergo its first hydroxylation, resulting in the formation of circulating vitamin D. The liver then transports vitamin D3 to the kidneys. From the liver it is transported to the kidney in the proximal convoluted tubule, where it is finally converted into calcitriol, its biologically active form (Holick, 1999).

Calcitriol is a hormone that regulates calcemia and bone mineralization through its actions on the intestine, bone, kidney and other tissues. It also promotes calcium absorption, as it acts on the enterocyte in the small intestine, where it stimulates the synthesis of the calcium-binding protein, calbindin (CaBP), which is necessary for the intracellular transport of calcium (Grudtner, 1997).

During food processing and digestion, inositol phosphate can be partially dephosphorylated and produces degradation products such as penta-, tetra- and triphosphate by the action of the enzyme phytase, present in the plant itself (Sanderberg et al., 1987).

Phytic acid has been considered an antinutrient due to its inhibitory effect on the bioavailability of minerals, as it has a chelating power with multivalent cations. The minerals most affected are calcium, iron, zinc and copper (Zhou, 1995). When phytic acid combines with calcium in the intestine, it forms an insoluble complex which cannot be absorbed and is therefore eliminated through the feces (Nordin, 1997).

The fibers present in cellulose, hemicellulose and lignins and found in large quantities in forage crops and legumes, seem to have a direct effect on the bioavailability of calcium. The negative effect of fibers on calcium absorption is mainly due to phytates, which are often associated with fibers (Guéguem et al., 2000).

Estimates of total daily calcium requirements are first made using endogenous losses and then for gain, gestation and lactation estimates, which are adjusted according to the absorption coefficient estimated by the NRC (NRC, 2007). However, to calculate calcium requirements for goats, the same formula as for sheep is used, due to the lack of data on endogenous losses, resulting in the following equation:

Calcium requirement for maintenance (Cam):

$$\text{Cam (g/dia)} = \frac{0{,}623 \times \text{CMS} + 0228}{0{,}40}$$

CMS = Dry matter consumption

The NRC (1981) also sets out requirements including animal activities, as shown in table 7.

Table 7. Calcium requirements (g) according to NRC (1981)

	Live weight	Mantenga	M+BA 25% inc	M+MA 50% inc	M+AA 75% inc
	10	1	1	1	2
	20	1	2	2	2
	30	2	2	3	3
	40	2	3	4	4
Ca	50	3	4	4	5
	60	3	4	5	6
	70	4	5	6	6
	80	4	5	6	7
	90	4	6	7	8
	100	5	6	7	8

M=maintenance; BA=low activity; MA=medium activity; AA=high activity; inc=increment.

Sousa (1998), who aimed to determine the Ca requirement of growing goats for maintenance using the endogenous fractions method and the comparative slaughter method, observed that the resulting estimate for Ca requirements using the endogenous fractions method was 31.02

mg/UTM/day. The author found a real absorption coefficient of 0.76 and was able to calculate the dietary Ca requirement for maintenance at 40.82 mg/UTM/day, when estimated using the endogenous fractions method. He also reported that the value found was lower than all the results obtained with sheep and cattle, and found no reference to results obtained with goats.

For the Ca requirement estimated by the comparative slaughter method, a value of 52.85 mg/UTM/day was found and concluded that the values found for the dietary calcium requirement for growing goats, obtained by the endogenous fractions method and the comparative slaughter method, were lower than those recommended by the NRC (1981) for goats and ARC (1980) and AFRC (1991) for sheep.

According to Meschy (2000), calcium requirements become complex due to the fact that excretion via feces and urine is considered to be very low, in addition to the complexity of the routes involved for this element.

The results found by Sousa (1998) corroborate the idea that there are differences in the requirements of goats compared to sheep and cattle, and also show the scarcity of information on mineral requirements for goats, which confirms the considerations reported by the NRC (2007), when it talks about the limitations found in estimating mineral requirements due to the lack of published work, especially in tropical regions.

Calcium for gain was estimated at 9.4 g/kg live weight by Meschy (2000). The calcium requirement for gain recommended by the NRC (2007) is 11g Ca/kg liveweight, because it considers that calcium excretion for goats is slightly lower when compared to sheep, which estimate their calcium requirement for gain between the range of 8.2 to 12.9 g/kg liveweight.

In view of the considerations reported by Meschy (2000), it can also be observed that, even in cattle, calcium requirements are also close to those found for sheep and goats, with values between 10.9 and 16.2 g/kg live weight according to the NRC (1997).

In a study carried out in Brazil, Nobrega (2009) working with ^ boer ^ sprd goats grazing in the semi-arid region, estimated the following calcium requirements for weight gain, shown in Table 8.

Table 8. Estimated net calcium requirements for weight gain in goats

Net requirements				
	Gain/day (g)			
BW (kg)	50	100	150	200
15	0,698	1,396	2,094	2,792
20	0,732	1,464	2,195	2,927
25	0,758	1,517	2,275	3,033
30	0,780	1,561	2,341	3,121

Adapted from Nobrega (2009).

The author concluded that net calcium requirements increased as the animals' body weight increased.

However, Souza (1998), estimating the calcium requirements for gain in growing Alpine goats, obtained the results shown in Table 9.

The author found lower values than those recommended by the ARC (1980) of 320 mg Ca/day for sheep, as well as the values recommended by the AFRC (1991) and similar to those recommended by the NRC (1981) and those found in Brazil by Resende (1989) and Ribeiro (1995).

Table 9. Calcium requirements for weight gain in growing Alpine goats

	Daily gain (g)		
Live weight	50	100	150
18	0,349	0,699	1,049
20	0,346	0,692	1,038
22	0,343	0,686	1,029
24	0,340	0,680	1,021
26	0,337	0,675	1,013

Adapted from Souza (1998).

However, he observed a drop in mineral requirements as the animals' body weight increased, which was not explained in this experiment.

When compared with the requirements found by the NRC (2007) table 10, it can be seen that the requirements found by the committee are always higher than the results found in the experiments conducted in Brazil:

Table 10. Calcium requirements for goat weight gain according to NRC (2007).

	Daily gain (g)		
Live weight	100	150	200

15	3,7	4,8	6,2
20	3,8	5,2	6,3
25	4,3	5,3	6,6
30	4,4	5,4	6,7

Adapted from NRC (2007).

However, it is known that there are differences in the results of studies conducted with animals in confinement and on pasture, since mineral requirements can be influenced by previous nutrition (NCR, 2000). This demonstrates the consistency of the reports and statements made by researchers regarding the differences in nutritional requirements between species, breed, physiological state, concentration and chemical form of the element in the diet, bioavailability, interrelationships with other nutrients and between nutrients (McDowell, 1996).

However, when looking at the body composition of calcium in the weight gain of goats, it can be seen that the retention of this element is dependent on the composition of the weight gain (bones, muscle and fat). However, when fat deposits are higher, there is a reduction in mineral deposition with a consequent reduction in the animals' requirements, since the concentration of these inorganic elements in adipose tissue is lower than in muscle and bone. Therefore, factors that modify the composition of the gain, such as the type of diet, sex, genetic group, age and weight of the animals, affect the mineral composition and, consequently, the net requirements for gain (Paulino et al., 1999).

The NRC (2007) reports that calcium requirements for pregnancy depend on the size and number of offspring.

However, during pregnancy, the most important phase in terms of nutritional levels is in the final third of pregnancy, comprising the last 50 days, when there is greater development of fetal tissues (Kadu and Kaikini, 1987). According to the NRC (2007), the concentration of calcium in the newborn fetus is estimated at 11.5 g/kg of fetal weight. It is also reported that 0.23 g/kg of fetal weight is required in the last 50 days of pregnancy (Table 11).

The ARC (1980) reports that in order to calculate the amounts of nutrients deposited in the uterus of pregnant goats at the various stages of pregnancy, it is necessary to know the amount of nutrients in the fetus, the additional amount deposited in the fetal membranes, fluids and uterine wall.

Table 11. Calcium requirements for pregnant goats according to the NRC (2007)

Requirements for pregnant goats with single offspring		
Live weight (kg)	Milk production (kg)	Ca (g/day)

20	2,3	3,5
30	2,9	3,8
40	3,4	4,2
50	3,8	4,4
Requirement for pregnant goats with double offspring		
20	2,1	4,9
30	2,6	5,5
40	3,0	5,8
50	3,8	6,2

Adapted from NRC (2007).

The idea is to get the animal to adequately mobilize bone calcium in order to sustain milk production at the start of lactation and maintain adequate blood levels, because if levels are insufficient at this stage, the animal will be at serious risk of aborting the calf. However, it is important that calcium levels in the diet are not high at the end of pregnancy.

Table 12 shows the Ca requirements for pregnant goats.

Table 12. Requirements of pregnant goats with two fetuses

Weight (kg)	Period	IMS (kg)	Ca (g)
40	Home	1,07	3,0
	4th month	1,07	5,0

	5th month	0,97	7,0
50	Home	1,20	3,5
	4th month	1,20	6,0
	5th month	1,09	8,5
60	Home	1,33	4,0
	4th month	1,33	7,0
	5th month	1,21	10,0
70	Home	1,47	4,5
	4th month	1,47	8,0
	5th month	1,34	11,5
80	Home	1,60	5,0

	4th month	1,60	9,0
	5th month	1,46	13,0

IMS = dry matter intake Adapted from Ribeiro (1997); NRC (1981)

For lactation, the NRC (2007) reports that the calcium required depends on the concentration of calcium in the milk and the milk yield, with the recommended use being 1.4 g of calcium per 1 kg of milk produced.

Alevinos (1911, 1913) cited by the NRC (1981), found that when lactating goats do not receive the necessary amounts of calcium and phosphorus in their diets, calcium is inevitably removed from the body without initially affecting milk production and composition, but can later cause a more severe negative calcium balance.

Chapter 2

Phosphorus (P) requirements

Phosphorus studies were given impetus by Brandt in 1669 when he managed to work with the element in isolation, finding its presence in human urine. Later, in 1769, Gahn presented reports on the presence of phosphorus in the composition of bones, emphasizing its essentiality. In 1771, the presence of phosphorus in bone tissue was observed by Scheele (Carvalho, 2005), and there have subsequently been several studies proving the importance of this mineral for animal health.

Later studies showed that the inorganic part of the bones was made up of calcium phosphate with a small part made up of carbonate and magnesium. Theiler, in 1920, clarified the cases described in 1785, where sheep on pasture had characteristics of botulism preceded by phosphorus deficiency (McDowell, 1992).

According to Andriguetto (1990), the animal body contains 0.9 to 1.1% phosphorus, with 16 to 17% represented in the ash. Of this total, around 80 to 85 % is located in the bones in the form of hydroxyapatite and the rest is part of the body's other tissues (Barcelos et al., 1998; Carvalho 2005).

According to Georgievskii et al. (1982), the body tissues with the highest phosphorus concentrations are the berry and the brain, with levels ranging from 350-400 mg/dl and 240-430 mg/dl, respectively. In the blood it appears in concentrations of 17-20 mg/dl and in the serum 3-7 mg/dl.

The least amount of phosphorus in the body is inorganic, i.e. in mineral form. According to Smith et al. (1952), sheep have variable P levels, representing: 14.3 in the liver, 11.0 in the kidneys, 9.0 in the heart and 6.7 mg/g MS in the muscle.

However, in soft tissues and body fluids, the organic form of phosphorus is predominant, forming compounds such as: phosphoproteins, hexose phosphates, creatine phosphate, among others. Also according to Runho et al (2001), it is involved in the formation of nucleic acids, energy transfer in the forms of adenosine mono-, di- and triphosphate, osmotic pressure and acid-base balance, etc.

However, it can be seen that phosphorus is a widely studied element that is part of many vital fungi in the body.

Barcellos et al. (1998) reported that phosphorus is available as inorganic mono-, di- and triphosphate and in organic form as phytate, phospholipids and phosphoproteins, which are absorbed in the small intestine. The absorption of dietary phosphate is 60 to 17% and is carried out by an active counter-transport system using sodium or simply by a passive diffusion process (Rosol and Capen, 1997).

In ruminants, most of the absorption takes place at the beginning of the duodenum. In some experiments it was possible to demonstrate that absorption not only took place in the duodenum, but also in the rumen by passive diffusion through the rumen epithelium (Rosol and Capen, 1997; Barcelos et al. 1998).

The absorbed P is quickly removed from the blood and fixed in the bones and teeth to be used in the dynamics of their metabolism, especially during the growth phase (Carvalho et al., 2003; Dayrell, 1986). The process of reabsorption and deposition is dynamic but balanced. During periods of high demand, such as pregnancy and lactation, increases in demand are high and the processes of reabsorption and deposition are accentuated (Andriguetto el al., 1990).

As with most nutrients, the absorption and recycling of P are greater in animals that are in phases of greater demand, be it growth, production or reproduction (Carvalho et al., 2003). The greater the requirement, the greater the animal's capacity to absorb phosphorus, and the same also occurs when there are insufficient sources of phosphorus in food. However, P absorption can be influenced by P sources (bioavailability), the age of the animals, and the presence of other nutrients such as lipids, calcium, iron, aluminum, potassium, magnesium and manganese, as well as being stimulated by vitamin D (Ammerman, 1967; Martin, 1993; Carvalheiro and Trindade, 1992).

Some substances can impair absorption when they are present in mixtures with P sources, such as high concentrations of magnesium and aluminum. However, phytate is highly available to ruminants due to the presence of rumen bacteria that make P available from plants (McDowell, 1992).

In addition to intestinal action, the reabsorption of P is also carried out in the renal tubules, as a way of adapting to the scarcity of P in the diet. However, absorption depends not only on its presence in the diet but also on the bioavailability of the P ingested (Rosol and Capen, 1997).

For ruminants, not only does phosphorus play an important role in the animal's body, but it is also very important for meeting the requirements of the rumen microflora, and if there is a deficiency of phosphorus in the rumen, microbial protein synthesis can be impaired (Ternouth & Sevilla, 1990).

There are high concentrations of phosphorus in the rumen, ranging from 200 to 600 mg/L (Witt & Owens, 1983), and around 50 to 70% of this phosphorus is endogenous, i.e. phosphorus secreted by saliva, which has two important functions: to act as a buffer against the low pH in the rumen, resulting from the production of organic acids, and to be part of the nutrition of the rumen microbiota. Microbial cells contain 20 to 60 g of P/kg of dry matter (Hungate, 1966), present as nucleic acids (80%) and phospholipids (10%).

However, ruminants are able to metabolize phosphorus more efficiently by secreting 16-40 mmol/l concentrates in their saliva, which increases the concentrates in the rumen and duodenum and facilitates absorption, which does not occur in non-ruminants (Rosol and Capen, 1997).

After absorption in the GIT and rumen, phosphorus circulates in two forms: part by complexing with plasma proteins and part in ionic form. It reaches the liver and is directed to the most diverse tissues. In ruminants, much of it goes into saliva.

According to Martin (1993), among the many functions that P performs in the animal organism, perhaps the most important is its participation in cell metabolism, as it is part of the biochemical mechanisms linked to energy metabolism.

However, P deficiency in animals raised on natural pastures is a reality. Even if a severe deficiency does not occur, the consequences on productive functions can be observed (Albiston, 1975).

Despite this, Barcellos et al. (1998), reports that during periods of deficiency, animals are able to remove up to 30% of the phosphorus deposited in the bones, which is first reabsorbed from the vertebrae and ribs, which are made up of spongy bones.

Low phosphorus diets cause changes in metabolism that allow the secretion of substances that promote the optimization of intestinal P absorption, such as vitamin D, which is responsible for increasing P absorption in the intestine.

The symptoms of phosphorus deficiency are: reduced weight gain, poor appetite, locomotion problems, low organ resistance and low fertility (Santos, 2000; Conrad and Souza, 1976; Barcellos et al., 2003).

MacDowell (1992) reports that many factors can affect phosphorus requirements such as: feed, stress, sex, environmental conditions, levels of other elements, growth rate, etc.

However, one of the difficulties encountered when determining phosphorus requirements is the recycling of this mineral from the blood to the saliva, which is very variable during the day, which is then taken to the rumen and ends up being added to the phosphorus in the diet, which can lead to an overestimation of the element coming from the diet.

Some work has been carried out in Brazil to obtain information to estimate nutritional requirements under Brazilian conditions, thus adapting basic knowledge from other countries to our reality (Silva, 1995).

It is well known, therefore, that younger animals are more demanding in terms of the levels of P in their feed, and their serum levels of inorganic P are usually higher than those of older animals.

Silva et al. (2002) using data from 14 experiments on nutritional requirements, using the comparative culling method, found that the nutritional requirements for phosphorus are higher than those recommended by the NRC (1996).

When estimating phosphorus requirements, the mineral deposition at different ages and weights should be assessed through chemical analysis of the tissues after the animal has been slaughtered. Using this information, models have been proposed to estimate the mineral composition of the animal's body and, consequently, the requirements for growth and gain (ARC, 1980; AFRC, 1991).

The NRC (2007) reports that the total phosphorus required for maintenance, growth,

pregnancy and lactation is also estimated using the factorial method proposed by the NRC, 1997), based on 138 tests, which led to the formation of the following equation for estimating phosphorus maintenance requirements:
Pm (g/day) = 0.081 + 0.88 x CMS

These values are lower than for sheep because it is reported that goats recycle phosphorus better through their saliva. The value for gain is 6.5 g P/kg of live weight gain.

For gestation requirements, the committee reports that it is necessary to know the number and size of the offspring. In this case, phosphorus is only corrected in the last 50 days of pregnancy, which is: 0.132 g/kg fetuses/day.

For lactation, the requirement is dependent on milk production, and is recommended within a range of 0.92 - 1.06 g/kg of milk, varying with the beginning and end of lactation, with a general recommendation of 1 g/kg of milk.

Bueno and Vitti (1999), estimating the phosphorus requirement for maintenance in Alpine goats, found a regression value of 10.36 mg/kg BW, which according to the NRC (1980), would be the minimum endogenous fecal loss of phosphorus. However, as the author found no concentration of phosphorus in the urine, the value could be considered the net requirement of the animals for maintenance, since the net requirement is the sum of the endogenous fecal loss plus the urinary loss. Using the absorption coefficient found of 65.76%, the net dietary requirement for maintenance can then be obtained, which was 15.75 mg/kg BW/day.

The NRC (1985) recommends 20 mg/kg BW as the net requirement for sheep. This corroborates what has been reported previously, where sheep tend to excrete higher concentrations of phosphorus than goats, because the latter have a greater recycling capacity.

Carvalho (2003), estimating the phosphorus requirements for Saanem goats, found a requirement of 6.87 mg/kg BW/day through the minimum endogenous fecal losses of 6.71 mg/kg BW/day added to the phosphorus found in urine of 6.87 mg/kg BW/day. The author concludes by emphasizing that phosphorus intake influences endogenous fecal losses of the element, which is also influenced by dry matter intake.

Queiroz (2000), also estimating the phosphorus requirements for maintenance of growing Alpine goats using the endogenous losses method and comparative slaughter, concluded that the daily dietary phosphorus requirements for maintenance were estimated at 72.0 mg/kg BW/day when estimated using the endogenous losses method and 58.46 mg/kg BW/day using the comparative slaughter method, resulting in 720.0 and 584.6 mg P/day using the endogenous losses method and comparative slaughter respectively.

The NRC (1981) recommends 700 mg/day for goats weighing between 20 and 30 kg of BW, which demonstrates the differences between the requirements of animals raised in temperate climates and those raised in tropical climates.

Nóbrega (2009) in a study with grazing goats concluded that net phosphorus requirements increased with increasing body weight, obtaining the values shown in Table 13.

Table 13. Estimated net phosphorus requirements for weight gain in goats.

Net requirements				
Gain/day (g)				
BW (kg)	50	100	150	200
15	0,328	0,655	0,983	1,331
20	0,329	0,658	0,987	1,316
25	0,330	0,660	0,990	1,320
30	0,331	0,662	0,992	1,323

Adapted from Nobrega (2009).

Carvalho et al. (2003) reports that the increase in the requirement for the element, to the detriment of the increase in body weight, is related to the rate of bone growth in the animals, indicating that the younger they are, the animals need increasing amounts of minerals for the structural development of the body.

Sodium (Na) and Chlorine (Cl) requirements

Sodium and chlorine are studied together because they are very important electrolytes in maintaining the osmotic pressure of extracellular and intracellular fluids and in maintaining the acid-base balance and because Na and Cl are associated in the form of common salts.

Underwood (1999), also shows this interaction between these minerals when he reports that it is convenient to study sodium and chlorine together because of their interactions, inferring the needs of the animal organism.

Sodium is mainly found in the extracellular fluid, with less than 10% inside the cell. Approximately half of the intracellular sodium is adsorbed to the hydroxyapatite crystals in bones. Blood plasma and most other extracellular fluids have a high sodium concentration.

Minson (1990) reports that sodium is the main cation in the extracellular fluid of ruminants and is required for microbial growth in the rumen.

Chlorine acts with bicarbonate to electrically balance the sodium in the extracellular fluid. It is found almost exclusively in the extracellular fluid and as an essential element of gastric juice.

The absorption of Na^+ and Cl^- takes place mainly in the upper portion of the small intestine and to a lesser extent in the lower portion of the small intestine and the large intestine.

Both ions are absorbed either actively or passively, depending on the site of absorption. Sodium crosses the mucosa by active transport in the intestine. Chlorine is absorbed by active transport in the upper portion of the small intestine, but by passive diffusion in the lower portion of the small intestine.

Plasma sodium levels tend to be controlled by the action of the hormone aldosterone, a mineralocorticoid produced by the cortex of the adrenal gland. It promotes the partial reabsorption of sodium in the renal tubules.

The production of aldosterone, in turn, is under the control of the adrenocorticotrophic hormone from the anterior pituitary gland. This antidiuretic hormone is released in response to changes in the osmotic pressure of the extracellular fluid induced by water deprivation. At the same time, the release of aldosterone is suspended and consequently there is renal reabsorption of sodium.

In the opposite situation, when water intake is high or there is a decrease in plasma sodium, the antidiuretic hormone is inhibited and aldosterone starts to act on urinary sodium reabsorption.

The concentration of chlorine ion in the extracellular fluid tends to balance out in relation to the concentration of sodium in the body. Excessive renal excretion of sodium ion raises the concentration of bicarbonate ion (HCO3-) so that an equal amount of chlorine ion (Cl-) is excreted through the urine. The relationship between these two ions is based on maintaining an identical concentration of cations and anions in the plasma.

Extracellular sodium and intracellular potassium act through a sodium-potassium pump to maintain osmotic balance. The concentration of potassium inside the cell is much higher than its concentration in the blood plasma, and the opposite is true for sodium. Maintaining these gradients across the plasma membrane depends on the supply of chemical energy from ATP. The cell membrane contains an enzyme called Na+ K+ transporting ATPase. It catalyzes the hydrolysis of ATP into ADP and organic phosphate and uses the energy released to pump K+ into and Na+ out of the cell, against the concentration gradient. For each ATP molecule hydrolyzed, three Na+ ions are transported out of the cell and two K+ ions into it. The Na+ K+ ATPase of the renal tubule cells works by allowing a constant loss of potassium in the urine, while the loss of sodium must be kept at very low levels (Dukes, 2006).

The Na+K+ pump is also responsible for nerve impulse transmission. It generates an electrical potential difference in the axonal membrane of the neuron, which is responsible for transmitting the nerve impulse along the neurons. Maintenance of water and acid-base balance. It is also a component of digestive secretions and participates in the transportation of nutrients and other electrolytes across cell membranes.

Chlorine acts to control extracellular osmotic pressure and maintain the body's acid-base balance. It is secreted in large quantities in the stomach to form HCl (gastric juice) together with the H+ ion.

Lately, researchers have focused more on determining the effect that the ionic balance can have on production results than on the absolute quantities of ions to be supplemented.

Chlorine deficiency, like potassium deficiency, is difficult to occur. The main signs are: decreased growth rate and the appearance of nervous symptoms, poor appearance of the skin, hair, decreased appetite, decreased conversion rate and production, and decreased appetite.

In general, eating more than you need results in it being rapidly excreted by the kidneys. As a result, it is unlikely that excessive sodium and chlorine in the feed will lead to intoxication. Except when water consumption is restricted, limiting urinary excretion, or when the animal has kidney failure.

Signs of poisoning are manifested by staggering gait, paralysis and convulsions. Excess chlorine produces a moderate increase in water consumption.

Drinking water can therefore be a valuable source of supplementary sodium in areas where food sodium intake is low, but it can provide excessive amounts of salt (and other minerals normally present with salt, such as sodium sulphates and bicarbonates, calcium and magnesium). When deprived of sodium, animals show a clear preference for salt water (Blair-West et al., 1968; Smith and Middleton, 1980). A similar preference has been observed in goats (Goatcher and Church, 1970).

ARC (1980) estimated the sodium requirements for maintenance in sheep considering 5.8 mg of fecal losses and 20.0 mg of urinary losses per kg live weight, using an absorption coefficient of 0.91%. However, sodium requirements are higher than those found in practice (NRC, 1985; Michell, 1995).

The NRC (2007) describes that sodium and chlorine requirements for goats are calculated using the factorial method for maintenance, growth, pregnancy and lactation using the formulas shown: Sodium:

Mantenga (g/day) = (0.015 x CP) / 0.80
Growth (g/day) = (1.6 x GPV) / 0.80
Gestation (g/day) = (0.034 x GPV) / 0.80
Lactation (g/day) = (0.4 x PL) / 0.80
Chlorine:
Mantenga (g/day) = (0.022 x PC) / 0.80
Growth (g/day) = (1.0 x GPV) / 0.80
Gestation (g/day) = (0.024 x GPV) / 0.80
Lactation (g/day) = (1.1 x PL) / 0.80

Nobrega (2009) emphasized the importance of using the sodium present in common salt, reporting that common salt (sodium chloride) is perhaps the mineral most commonly supplemented to animals and when offered freely, where goats can consume salt above their requirements, but without any apparent harmful effects. In his research, the author found the sodium requirements shown in Table 14.

Table 14. Estimated net sodium requirements for weight gain in goats

	Net requirements			
	Gain/day (g)			
BW (kg)	50	100	150	200
15	0,058	0,116	0,174	0,232

20	0,067	0,135	0,202	0,270
25	0,075	0,151	0,226	0,302
30	0,083	0,165	0,248	0,330

Adapted from Nobrega (2009).

The author reports that the results found in his work had values above

(1996), who reported requirements of 0.88 to 0.86 mg of Na per kg of gain for animals weighing 15 to 25 kg of BW, and Fernandes (2006), who found values of 0.57 to 0.54 mg of Na per g of gain for animals weighing 20 to 30 kg of BW.

Geraseev et al. (2001) estimated the requirements of Santa Inês lambs and found the values shown in Table 15.

Table 15. Estimated net and dietary sodium requirements for maintenance and live weight gain (g/animal/day)

	Live weight (kg)	Maintenance (g)*	100	200	300
Liquid	25	0,646	0,063	0,126	0,189
	30	0,773	0,059	0,118	0,177
	35	0,901	0,056	0,112	0,168
Dietetics	25	0,71	0,069	0,138	0,207
	30	0,85	0,065	0,130	0,195

	35	0,99	0,061	0,122	0,183

*Value recommended by ARC (1980) Adapted from Geraseev et al. (2001).

In this study, the authors observed that the values found were 74.6% for lambs weighing 25 kg and 96.4% for lambs weighing 35 kg when compared to the NRC (1980). The reasons given for this disparity could be due to the differences in the body composition of the animals and, above all, the influence of climatic conditions, which is very relevant given that in warmer climatic conditions there is certainly a loss of sodium through transpiration.

Underwood (1999) reports that sodium and chlorine are lost through skin secretions, but emphasizes that this can vary greatly between animal species.

Costa et al. (2005), working with goat retention during pregnancy, where they compared Na retention in the pregnant uterus + udder and the estimate through a regression model in goats with different stages of pregnancy, can observe the results found in table 16.

Table 16. Comparison between sodium retention in the pregnant uterus + udder (Y) and its estimation using the regression model lnY= A + Bx + Cx2 (Y1) in goats at <u>different stages of pregnancy, with one and two foetuses.</u>

Length of pregnancy/No. of fetuses	Estimate	Na (mg)
50 days/1 fetus	Y	0,89
	Y1	1,06
100 days/1 fetus	Y	6,50
	Y1	6,46
140 days/1 fetus	Y	13,18
	Y1	13,19
50 days/2 fetuses	Y	1,23

	Y1	1,23
100 days/2 fetuses	Y	10,84
	Y1	10,84
140 days/2 fetuses	Y	21,44
	Y1	21,44

Adapted from Costa (2005).

The authors observed that at 50 days of gestation, the model overestimated Na retention for goats with one fetus, which was not observed for goats with two fetuses, and report that, however, as the first third of gestation is the phase in which, quantitatively, mineral retention is minimal, the use of any of the estimates would not cause considerable damage.

They also observed that for the period from 100 to 140 days of gestation, mineral deposition for pregnancies with one and two fetuses represented 50.7 and 49.4% (Na) of the total deposited up to 140 days in the products of gestation, concluding that the estimates made by the exponential polynomial model $lnY = A + Bx + Cx2$ efficiently represented the biological behavior of mineral retention during the gestation of goats.

Chapter 3

Potassium requirements (K)

In Latin, potassium is called Kalium. According to Tisdale et al. (1993), in the soil, K can be present in four categories in terms of its availability: structural, non-exchangeable (where it is not available to plants), exchangeable, and in solution. When these forms are added together, we have total K.

According to Sparks (1980), the passage of exchangeable K into the soil solution is very rapid, so that K is highly available to plants through this exchange. Non-exchangeable K is moderately available, since its passage into the exchangeable form is slower.

In plants, the availability of K depends, among other factors, on the amount of exchangeable K, the concentration of K in solution and the relationship between them, thus determining the capacity of soils to maintain a concentration of K available for absorption by plants. However, according to Sparks (1980), these fragments are in dynamic equilibrium. They may or may not be at the ideal concentrations to meet the plants' requirements for K. However, K is considered to be abundant in nature, with between 1 and 4% found in fodder and an average of 0.5% in grains.

However, until recently it was believed that K deficiency was difficult to occur, as young forages generally contain considerably more K than is required by the animal. However, it has been suggested that in high-producing ruminants, K requirements can reach over 10 g/kg under stress, particularly heat stress (McDowell, 1985; Khan et al., 2005). Similar K concentrations were also reported by Prabowo et al. (1990) in Indonesia, Ogebe et al. (1995) in Nigeria and Tiffany et al. (2000, 2001) in northern Florida.

This can probably be influenced by using mature pastures in winter, which are exposed to bad weather and hay that is exposed to a lot of rain and sun, which can interfere with K levels.

Souza et al. (1982), in a survey of mineral deficiencies in cattle in the north of Mato Grosso, found average levels of K in forage crops that were adequate for the nutritional requirements of cattle.

However, Sousa et al. (1987), carrying out a survey in the state of Roraíma, found that K levels in forage crops were deficient in all regions, and were significantly lower in the dry season.

Khan (2009), evaluating the mineral composition of forages for grazing ruminants in Pakistan, observed that the levels of K were mostly always high, as shown in graph 1.

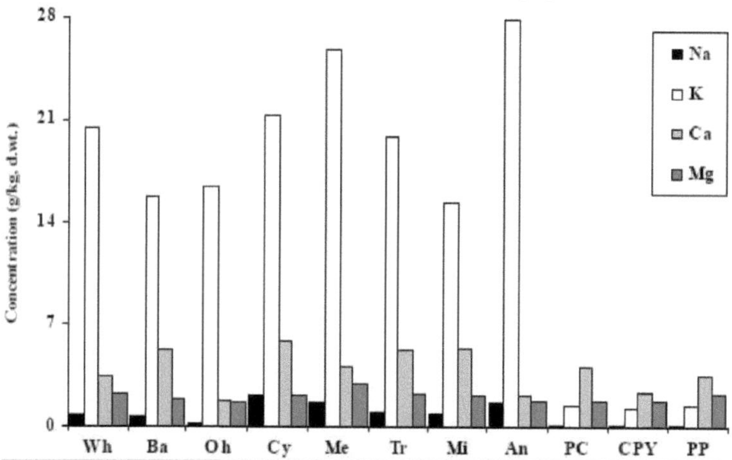

Graph. 1. macromineral content of some dry leaves from different feedstuffs [Wheat (Wh), barley (Ba), oat hay (OH), Cynodon (CY), Medicago (Me), Trifolium (Tr), Mista (Mi), Andropogon (An), Paspalum cojugatum (PC), Cynodon plectostachchyum (CPY), Pennisetum purporeum (PP)].
Adapted from Khan (2009).

Depending on the levels of K in the pasture and/or in the ingredients used in the feed, K can be supplemented with various chemical forms such as: chlorides, carbonates, bicarbonates and orthophosphates, which are of approximately equal biological value.

Carvalho (2005) states that more than 95% of the K ingested by ruminants raised on pasture is metabolized and absorbed by simple diffusion, with the absorption sites being found in the rumen, omasum and final portion of the small intestine. Its recycling is high and one of the indications of this is the presence of high concentrations of the element in saliva, which is the main source of K for rumen microorganisms.

According to Thompson (1972), K is considered to be a useful and non-critical nutrient, as it is involved in several important functions in animal metabolism, such as: regulation of osmotic pressure, water balance, regulation of rumen pH, which is fundamental for promoting an ideal environment for rumen microorganisms. It is also important, together with sodium and chlorine, in maintaining ionic balance and acid-base balance, is extremely important in the transport of O2 and CO2 in the blood and is important in the transmission of nerve impulses and contractility in muscle fibers.

The plasma level of K also tends to be controlled by the action of the hormone aldosterone, a mineralocorticoid produced by the cortex of the adrenal gland. It is responsible for the partial reabsorption of potassium by the kidneys.

The main route of potassium elimination is through urine, but it can also be eliminated through sweating, which occurs on the skin, and secretions from the upper airways. In the feces, the percentage of K lost is around 13%. This loss in the feces can be increased when there is a problem such as inflammation of the intestines, which causes diarrhea and a greater loss of the mineral.

Potassium consumption by animals generally exceeds the animal's metabolic requirements. However, even with this condition, potassium intoxication is difficult to occur due to the kidney's ability to control its excretion. Aldosterone also plays a role in maintaining potassium in the body. A high concentration of potassium in the extracellular fluid stimulates the secretion of aldosterone in the same way as a low concentration of plasma sodium. In a potassium deficiency, part of the sodium is transferred to the inside of the cell to replace the potassium, thus maintaining the osmotic and acid-base balance.

The NRC (1981) reports that K is generally required in large quantities, but this does not become a problem due to the fact that there is good availability in forage and grain. It also reports that deficiencies are related to a reduction in food intake, causing a delay in growth and milk production. The committee also reports that in sheep, the requirement for growth is considered to be 0.5% of the diet provided.

The NRC (1989) reports that the coefficients can vary from 91 to 100%. The requirement for K ranges from 0.5 to 0.8%, and seems to be higher for animals under stress. The understanding is that, with excitement, there is a tendency to increase the loss of this element through urine, with diseases and diarrhea being a cause of this type of loss.

However, the NRC (2007) presents some formulas for estimating K requirements:
Maintenance:
K mantenga (g/day) = (2.6 x CMS + 0.05 x PC) / 0.90
Growth:
K growth (g/day) = (2.4 x ADG) / 0.90
Gestation:
K gestation (g/day) = 0.042 x LBW) / 0.90
Laceration:
K lactation (g/day) = 2.0 x MY) / 0.90

Mcdowell (2000) warned that in dry periods there is a possibility of K deficiency, as well as in mature pastures, in hay made from pastures that are exposed to a lot of sun and rain, or in animals fed on grain.

In the work carried out by Nobrega et al. (2009), the author reported that the net requirement obtained for potassium in his experiment (Table 17) was similar to that obtained by Resende et al. (1996) and Fernandes (2006), who obtained, respectively, net requirements for this mineral ranging from 1.01 to 0.95 mg, considering animals weighing 15 to 25 kg and 1.01 to 0.93 mg g-1 of gain, considering animals weighing 20 to 30 kg of BW. Table 17. Estimated net potassium requirements for weight gain in goats

	Net requirements			
	Gain/day (g)			
BW (kg)	50	100	150	200
15	0,049	0,098	0,147	0,196
20	0,049	0,098	0,147	0,196
25	0,049	0,098	0,147	0,196
30	0,049	0,098	0,147	0,196

Adapted from Nobrega (2009).

Geraseev et al. (2001) estimated the net and dietary K requirements of lambs.
Santa Ines, obtained the following results (table 18).

Table 18. Estimated net and dietary potassium requirements for maintenance and live weight gain (g/animal/day)

Live weight (kg)	Mantenga (g)*	100	200	300
25	2,112	0,200	0,400	0,600
30	2,435	0,192	0,384	0,576
35	2,657	0,186	0,372	0,558

*Value recommended by ARC (1980) Adapted from Geraseev et al. (2001).

Comparing the results found for goats and sheep, it can be seen that despite the fact that the data is sometimes used together, it is clear that the requirements for these two species are different.

And this, considering that all the experiments mentioned were carried out in a tropical climate, corroborates the considerations made earlier, when it is reported that, in addition to the environmental conditions that can interfere with mineral requirements, other factors such as breed, sex, physiological condition, among others, can have a strong influence on the mineral requirements of animals.

These considerations can also be emphasized when, for example, we look at the experiment carried out by Costa et al (2003), who estimated K requirements for goats with one and two fetuses, and found differences between the net K requirements obtained for goats with single and twin pregnancies, as shown in tables 19 and 20.

Table 19. Estimated net daily K requirement in pregnant goats with one and two fetuses.

Number of fetuses	Number of days	K (g)
	50	0,0315
	75	0,0865
1 Fetus	100	0,1690
	120	0,2185
	140	0,2052
	50	0,0431
	75	0,1348
2 Ferns	100	0,2809
	120	0,3610
	140	0,3090

Adapted from Costa (2003).

Table 20 - Estimated daily dietary potassium requirement in pregnant goats with one and two fetuses

Number of fetuses	Number of days	K (g)
	50	0,0315
	75	0,0865
1 Fetus	100	0,1690
	120	0,2185
	140	0,2052
	50	0,0431
	75	0,1348
2 Ferns	100	0,2809
	120	0,3610
	140	0,3090

Adapted from Costa (2003).

Magnesium (Mg) requirements

Magnesium (Mg) is considered an abundant mineral in most foods. It is more available in legumes than in grasses and in concentrates than in roughage. It has a high absorption coefficient, which decreases with age. True absorption in adult ruminants fed hay or forage ranges from 10 to 37% according to (ARC, 1980). However, research carried out in Brazil has shown a wide variation in the absorption coefficient, ranging from 16.3% (Coelho Da Silva et al., 1991), 43.7% (Rosado, 1991) and 56.9% (Valadares Filho et al., 1991).

Bones contain 0.8% Mg, corresponding to around 60 to 70% of the magnesium present in the body. According to Cardoso (2006), it is an important constituent of bones and teeth, cell membranes and chromosomes. The remaining 30 to 40% is distributed in soft tissues and body fluids.

In soft tissues, it is found in greatest concentration in the liver and skeletal muscles. Within soft tissue cells, magnesium is found in higher concentrations than any other element except potassium. The small amount present in the extracellular fluid is easily exchanged with that adsorbed on the surface of bone. The serum magnesium level normally varies between 1 and 3 mg/dl.

It is vital for the metabolism of carbohydrates and lipids, as a catalyst for a wide variety of enzymes (Underwood & Suttle, 1999).

In glycolysis, it is present in almost all the reactions involved in the conversion of glucose into pyruvate. Mg ions inhibit the release of acetylcholine. It is of great importance for muscle relaxation; when the concentration of Mg in the fluid that bathes the motor plate of the muscle fibers decreases, symptoms of tetany appear; at the same time, excitability disorders occur in the synapses activated by acetylcholine. The levels of Ca and P, and probably K, influence the animal's need for Mg. In the blood, it is distributed approximately 75% in the red cells and 25% in the serum.

It also participates in neuromuscular transmission and is necessary for potassium transport and calcium activity. Magnesium is known as a "natural calcium channel blocker". When magnesium is depleted, intracellular calcium rises. Since calcium plays an important role in the contraction of both smooth and skeletal muscles, magnesium depletion can result in muscle cramps, hypertension and coronary and cerebral vasospasms.

Magnesium is an intracellular mineral that also plays a fundamental role in various biological reactions. It is an activator of enzyme systems that control the metabolism of carbohydrates, lipids, proteins and electrolytes; it influences the integrity and transport of the cell membrane; it mediates muscle contractions and nerve impulse transmissions and is a co-factor in oxidative phosphorylation. Magnesium is essential for the fixation of calcium in bones and can cause or aggravate osteopenia and osteoporosis in adult animals and hinder the correct calcification of bones in young animals.

Mg's main function is to stabilize the ATP structure in muscle and other soft tissues. The substrate for the enzymes that use ATP is the Mg-ATP complex. In intracellular fluid containing high concentrations of Mg+, ATP and ADP exist mainly as MgATP-2 and MgADP- complexes. In many enzymatic reactions in which ATP participates as a phosphate donor, its active form is actually the MgATP-2 complex.

According to the NRC (2007), magnesium is part of more than 300 metabolic reactions for energy production and forms a complex with adenosine triphosphate to supply energy.

These include the enzymes hexokinase (Figure 3) and glucokinase in which Mg is a cofactor and which act in the phosphorylation of glucose in the cytosol in the preparatory phase of glycolysis, as well as glucose-6-phosphatase, phosphoglycerate kinase, pyruvate kinase and enolase (Figure 4) which act at different points in the glucose in the cytosol.

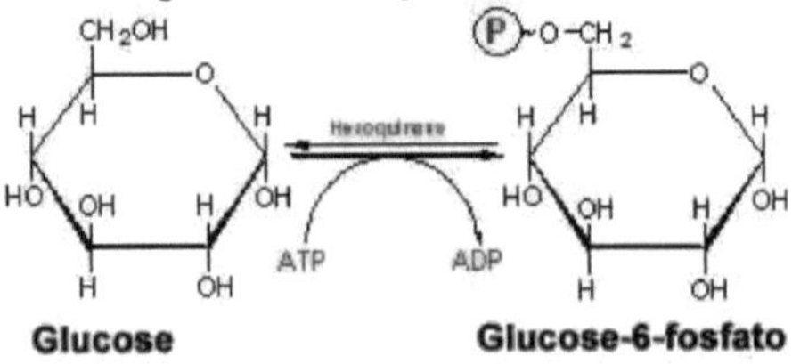

Figure 3 Transfer reaction of the external phosphoryl radical from ATP to glucose, catalyzed by exokinase with Mg as a cofactor.

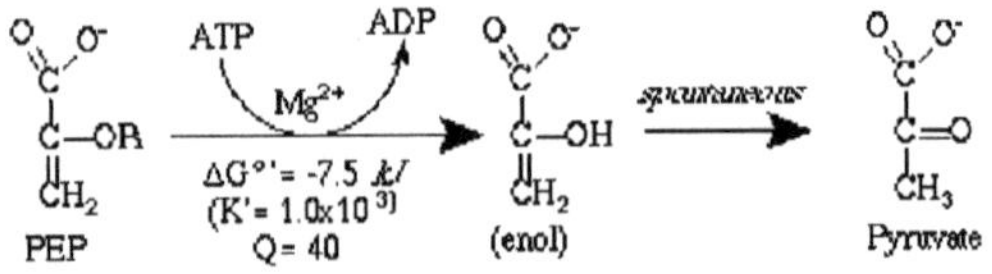

Figure 4 - Effect of enolase on the glycolysis pathway in the cytosol, with Mg as a cofactor.

Mg also acts as a cofactor for different enzymes in the oxidation of glucose via the pentose-phosphate pathway. It is a cofactor for the enzymes glucose-6-phosphate dehydrogenase, which transforms glucose-6-phosphate into 6-phosphoglycone-lactone, and lactonase, which transforms the latter metabolite into 6-phosphoglyconate. 6-phosphogluconate dehydrogenase, which transforms this into ribulose-5-phosphate, and thiamine pyrophosphate transketolase, which transforms both ribose-5-phosphate and glyceraldehyde-3-phosphate in the pentose-phosphate cycles.

It is also an essential activator for enzymes that transfer the phosphate group with myokinase which is important in the adenylate system of ATP formation where: ATP + AMP by action of myokinase form two molecules of ADP, which by oxidative ionization with 2 Pi produce two ATP in addition to difoffopiridinadinucleotide kinase and creatine kinase.

Another very important function is its participation in the metabolism of calcium, potassium, phosphorus, zinc, copper, sodium, hydrochloric acid, acetylcholine, nitric oxide, enzymes, intracellular homeostasis and thiamine activation. In the blood, it is distributed approximately 75% in the red cells and 25% in the serum.

According to the NRC (2007), the sites of Mg absorption occur primarily in the small intestine and rumen-reticulum in young and adult animals respectively

The NRC (1996); Grace et al. (1974) also report that the main site of Mg absorption in ruminants is in the rumen.

Tomas and Potter (1976) observed that magnesium infused into the omasum or abomasum of adult sheep, aged 2 to 5 years and weighing 40 kg live, was completely recovered in the duodenum. In contrast, 36 to 61% of the magnesium infused into the rumen was not recovered in the duodenum, suggesting that substantial magnesium absorption occurred in the rumen-reticulum. This suggests that post-rumen absorption may lead to an insufficient supply of Mg to the body.

The transport of Mg in the rumen mucosa occurs by active transport linked to Na, which is critical if the concentration of Mg in the diet is low. Therefore, adding Na to the diet can improve Mg absorption and transport. Unsaturated fatty acids in fodder can form insoluble complexes with Mg, compromising its ruminal absorption.

The homeostatic control of magnesium in the blood and tissues is unclear. Apparently, this is controlled by the excretion mechanism. Approximately 95% of magnesium is reabsorbed in the kidneys, the rest is eliminated through urinary excretion.

Urinary excretion is generally a reflection of the amount of magnesium absorbed. It has been suggested that, in ruminants, magnesium absorbed in excess of nutritional requirements is excreted via urine (Rook and Storry, 1962). Chicco et al. (1972) reported a high correlation ($r = 0.95$) between magnesium absorption and urinary excretion.

Mg is actively reabsorbed in the nephron and passively in the proximal tubule. Urinary (1.4mg/Kg/day) and fecal (0.5mg/Kg/day) excretion occurs, with the endogenous fecal fraction being excreted mainly in the proximal portion of the small intestine.

The kidneys conserve magnesium efficiently, particularly when its intake is low. The effects of magnesium can be mediated through its action as a calcium antagonist or by being a co-factor in enzyme systems that involve the flow of sodium and potassium across the cell membrane. As a result, smooth muscle relaxation occurs, cholinergic neuromuscular transmission is inhibited and mast cells are stabilized.

The action of thyroid hormones, acidosis, aldosterone and phosphate and potassium depletion increase magnesium excretion. On the other hand, calcitonin, glucagon and parathyroid hormone increase the reabsorption of glomerular filtrate (Cardoso, 2006).

The measurement of serum, plasma or erythrocyte magnesium is the most widely used indicator of nutritional status. Urinary magnesium excretion is used to assess magnesium nutritional status with an overload test, which is considered the most reliable method for detecting magnesium deficiency (Bohl and Volpe, 2002). As these are the tissues with the highest concentration of magnesium in the body, determining the content of this mineral in bone and muscle effectively reflects their body reserves. However, the techniques used to obtain tissue samples from muscle and bone are highly invasive and limit research (Bohl and Volpe, 2002).

According to the NRC (2007), dietary magnesium requirements for sheep and goats can vary with age, growth rate or production. The recommendations made by the NRC (1975) presented requirements of 0.040.08 percent of dry matter in the diet, which were later increased to 0.12, 0.15 and 0.18 percent of dry matter for growing lambs, late gestation females, and early lactation ewes, respectively by the NRC (1985). In addition, an increase in dietary magnesium of 0.20% was recommended for lactating ewes on pastures with a high nitrogen and potassium content. It also considers the absorption coefficient of 0.17 for sheep (ARC, 1980) and 0.20 for goats (Kessler, 1991), providing margins to avoid pasture tetany.

However, this type of deficiency is not common in stabled animals, but in grazing animals there is the so-called pasture tetany or hypomagnesemia. Thomas (1965); Rayssiguier et al. (1977) emphasize that the disorder is more common in older animals than in young animals, because magnesium is less labile.

Pasture tetany usually occurs during the early spring or in the particularly wet fall, when animals feed on fodder and grains in winter. It manifests itself clinically through stunted growth, hyperirritability, lack of appetite, incoordination and muscle convulsions, which can be followed by coma and death unless appropriate therapeutic measures are taken (Underwood & Suttle, 1999).

According to Larvor (1976), tetany seems to be caused by a physiological deficiency of magnesium, which can result from a simple deficiency in the diet or due to a low efficiency of utilization of the element. However, it is reported that there is some evidence that a displacement of the magnesium ion in the body is responsible, at least in part, for hypomagnesemia. Duncan et al. (1935) pointed out that perhaps the main reason why ruminants are more susceptible to hypomagnesemia than non-ruminants is because of the bioavailability of magnesium in the concentrates and roughage offered to these animals.

The recommendations presented by the NRC (2007) for magnesium requirements were estimated by the factorial method, using available data from seven studies and recommendations from the ARC (1980) and AFRC (1997). The magnesium requirements for pregnant goats were calculated for the last 50 days of pregnancy.

The formulas presented by the NRC (2007) for calculating requirements in goats are:

Mg maintenance (g/day) = (0.0035 x BW) / 0.20

Mg growth (g/day) = (0.40 x ADG) / 0.20

Mg gestation (g/day) = (0.006 x LBW) / 0.20

Mg lactation (g/day) = (0.14 x MY) / 0.20

Alcaide et al. (1999) describe that the use of Mg sources for supplementation in goats is based on the results found for cattle and sheep, with no specific data for goats. They also emphasize that among the other minerals, Mg is one of the most under-researched.

Reporting on the importance of using Mg absorption coefficients to estimate the total requirements for the element, the author carried out a study to determine the apparent absorption of Mg and found averages of 57.8; 73.9; 73.2% of absorption coefficient for the levels of 0.05; 0.20 and 0.35% of Mg DM in the supplement respectively. An interaction was also found for the real absorption coefficient in relation to the Anglonubiana and Saanen breeds, with the following averages: 61.0; 77.2; 73.2% for the Anglonubiana breed and 73.3; 75.5 and 76.0 for the Saanen breed in relation to the levels of 0.05; 0.20 and 0.35% of Mg in the DM.

The type of study carried out by Alcaide et al. (1999) is relevant because in the future it could be used to establish the value of the absorption coefficient in goats under Brazilian conditions, which is used to calculate the total net Mg requirements for animals in tropical climates.

Gerassev (2001), estimating the Mg requirements of Santa Ines lambs from 25 to 35 kg live weight, presented the results shown in Table 21.

Table 21. Estimated net and dietary magnesium requirements for maintenance <u>and live weight gain</u> (g/animal/day)

	Live weight (kg)	Mantenga (g)*	100	200	300
Liquid	25	0,075	0,039	0,078	0,117
	30	0,090	0,037	0,075	0,112
	35	0,105	0,036	0,072	0,108
Dietetics	25	0,441	0,229	0,459	0,687
	30	0,529	0,218	0,435	0,653
	35	0,618	0,212	0,423	0,635

*Value recommended by ARC (1980) Adapted from Geraseev et al. (2001).

    The author concluded that based on the results obtained in his experiment, the requirements presented in the tables developed with different rags and climatic conditions do not reflect the real Mg requirements of Santa Inés lambs reared in Brazil.

This conclusion reported by the author only corroborates the various considerations observed in the vast majority of nutritional requirements studies carried out in Brazil. The work carried out by Costa (2003) aimed to estimate the requirements of goats at different stages of pregnancy with one and two fetuses and obtained the results shown in tables 22 and 23.

Table 22 - Estimated net daily magnesium requirement in pregnant goats with one and two foetuses

Number of fetuses	Number of days	Mg (g)
	50	0,0030
	75	0,0089
1 Fetus	100	0,0219
	120	0,0392
	140	0,0610
	50	0,0038
	75	0,0133
2 Ferns	100	0,0367
	120	0,0695
	140	0,1118

Adapted from Costa (2003).

    The author observed that there was a difference in Mg requirements for goats with one and two fetuses and also between the data obtained for goats and sheep and also emphasizes that more studies should be carried out on mineral requirements for the goat breed.

Table 23. Estimated daily dietary requirement for magnesium in pregnant goats with one and two fetuses.

Number of fetuses	Number of days	Mg (g)
	50	0,0150
	75	0,0445
1 Fetus	100	0,1095
	120	0,1960
	140	0,3050

	50	0,0190
	75	0,0665
2 Ferns	100	0,1835
	120	0,3475
	140	0,5590
	140	0,3090

Adapted from Costa (2003).

Chapter 4

Sulfur (S) requirements

Sulphur is considered one of the most abundant elements in nature and is an essential element, particularly for ruminants (Carvalho, 2005).

Sulphur is an important element in protein synthesis, as it forms part of the molecular structure of essential amino acids such as: methionine, cystine, cysteine, homocysteine, cystathionine, taurine and cysteic acid, thiamine, biotin, lipoic acid, coenzyme A, glutathione, chondroitin sulphate, fibrinogen, heparin, ergothionine, and estrogens. Therefore, it is expected that most diets that provide adequate concentrations of protein will also provide adequate amounts of sulphur. Similarly, S forms part of some vitamins such as thiamine and biotin.

Sulphur in the mammalian body represents around 0.15 percent of body weight. Fortunately, all of the above sulphur compounds, except thiamine and biotin, can be synthesized in vivo from an essential amino acid, methionine.

Approximately 50% of the total requirement for sulphur-containing amino acids can be supplied by cystine.

According to the NRC (1984), sulphur is present in fodder as inorganic ionic sulphate and in a variety of organic compounds, mainly as sulphydryl groups and their derivatives. It is also found as chondroitin sulfate, an important component of cartilage, bones, tendons and blood vessel walls.

The body functions in which sulphur is involved, in addition to protein metabolism, are lipid and carbohydrate metabolism, blood coagulation, endocrine functions and the acid-base balance of intra- and extracellular fluids.

Sulphur is widely found in food as a constituent of protein, however deficiency in the soil can reflect on its content in the plant, which is directly related to the protein content of the food.

The organic forms of sulphur are easily absorbed through active transport and take place in the small intestine, especially the ileum, while the absorption of inorganic sulphate from the digestive tract is not very efficient. Its absorption must be constant, as the animal has little ability to store it. Trials with rats have shown that 41 to 64% of the inorganic sulphate in the diet is eliminated in the urine in no more than 8 hours.

Inorganic sulfate is excreted through the feces and urine. Sulphur that is not absorbed is probably reduced in the lower digestive tract as sulphate and endogenous faecal sulphur reaches the digestive tract mainly via the bile.

Urinary sulphur is found mainly as inorganic sulphur, but also as other organic components (thiosulphate, taurine, cystine, etc.). Most of the body's sulphur is found in proteins and so urinary sulphur excretion tends to be associated with urinary nitrogen excretion.

Sulphur is considered to be the best source of methionine for animals, but it must be added to feed as an amino acid and therefore as a precursor to the protein molecule. Economically viable sources include calcium sulphate, sodium sulphate and potassium sulphate.

Sulphur metabolism differs markedly between monogastrics and ruminants, and an understanding of this difference is fundamental for an appreciation of the sulphur cycle (Postgate, 1968) and the nutritional value of sulphur compounds. The main terrestrial source of sulphur is mineral sulphide, which is converted to inorganic sulphate by weathering and to organic sulphur by microbial action in the soil (Young & Maw, 1958). Inorganic sulphate is absorbed by plants and converted into organic sulphur in the form of sulphur containing amino acids, which in turn serve as a source of organic sulphur for ruminants. Many rumen bacteria are also capable of converting inorganic sulphur into organic sulphur in the form of methionine, cysteine and cystine and, consequently, into the various sulphur fungi in the body.

Rumen microorganisms need sulphur for the performance of their normal functions, without which there is a decrease in the digestibility of food and less nitrogen retention. They can incorporate inorganic sulphur (from forage and mineralized salts) into organic compounds and this is used for the synthesis of sulphur amino acids which are incorporated into microbial protein. As such, there is a Nitrogen:Sulphur (N:S) ratio considered to be optimal at 15:1 (varying according to the literature between 9 and 16:1) so that rumen microorganisms can synthesize sulphur amino acids from non-protein nitrogen (NNP) sources.

Table 24 shows the concentrations of amino acids in the cells of bacteria with different levels of sulphur.

In the case of sulphur deficiency, microbial protein synthesis is reduced and the animals show signs of protein malnutrition. Lack of S allows the development of microorganisms in the rumen that do not use lactate, and weight loss, weakness, tearing, dizziness and death can occur. A good N/S ratio must be considered in order to maximize the activity of microorganisms in the rumen, and S requirements for cellulose digestion are higher than for starch digestion (Pedreira & Berchielli, 2006).

This use is possible because rumen bacteria have the ability to reduce sulphate to sulphite. This reduction occurs at a pH of around 6.5. The incorporation of sulphate into cysteine is faster than into methionine, and utilization is faster when the source of S comes from concentrate, compared to forage (Church, 1993).

Table 24. Amino acid concentrations (mg / g DM) in rumen fluid bacterial cells from goats differing in the concentration of sulphur in the diet.

Amino acid _	% sulfur in the diet			
	0,11	0,20	0,28	0,38
Essential				
Arginine	13,28	13,91	13,73	13,57
Histidine	5,16	5,33	5,37	5,05
Isoleucine	13,79	13,71	14,06	13,81
Leucine	21,63	22,16	22,39	22,24
Lysine	19,36	19,71	20,46	18,97

Methionine	9,84	9,84	10,02	9,56
Phenylalanine	13,67	14,06	13,74	13,61
Threonine	14,28	14,68	14,29	14,11
Valine	15,39	15,62	15,59	15,30
Total	126,40	129,02	129,65	126,22
Not essential				
Alanine	23,83	23,94	23,51	23,41
Aspartate	32,09	33,20	32,94	32,34
Cysteine	3,28	3,77	3,80	3,65
Glutamate	38,16	40,59	39,55	38,72
Glycine	16,79	16,62	16,91	16,53

Proline	7,34	7,75	7,80	7,72
Cerina	12,81	13,14	14,35	13,71
Tyrosine	21,01	20,99	23,97	22,27
total	155,31	160,10	162,83	158,35

Adapted from Carneiro (2000).

The content of S in pasture depends largely on its concentration in proteins and rarely exceeds 3g/kg of DM. Methionine and cysteine generally contain more than 90% of the organic sulphur in forage (Gooneratne et al., 1989). However, sulphur varies widely due to alternative metabolic pathways in the rumen.

Sulphur leaves the rumen mainly through absorption as sulphide (S2-), but also through excretion in the form of protein sulphur or microbial protein. Only degraded protein sulphur and inorganic S from the diet or saliva are available to interact with Mo and Cu in the rumen. Diets rich in sulphur but low in Mo increase the rumen concentration of sulphide and decrease the availability of Cu in sheep (Suttle, 1991).

Sulphur supplementation is probably necessary to meet the requirements of ruminants when low-quality forage is grown on sulphur-deficient soils or when feed combined with non-protein nitrogen (NNP), such as urea, is provided.

Sulphur poisoning is not a common problem because the intestinal absorption of inorganic sulphur is very low. Sulphur is considered to be one of the least toxic elements.

Toxicity depends on the animal's ability to form H2S from inorganic SO4 sources, as H2S competes with cyanide. This can happen when animals are supplemented with substances containing high concentrations of sulphur (ammonium sulphate) in diets where there is a lot of NNP.

In the case of sulphur deficiency, microbial protein synthesis is reduced and the animals look dystrophic. The lack of sulphur allows the development of microorganisms in the rumen that do not use lactate, which can accumulate in the rumen, blood and urine.

With the growing use of non-protein nitrogen (NNP) to supplement part of the protein in ruminant diets, the likelihood of sulphur deficiency is increasing.

Sulphur supplementation can be important for ruminants fed low-quality roughage produced on sulphur-poor soils or roughage supplied with some source of non-protein nitrogen. The maximum tolerable concentration of sulphur in ruminant diets has been estimated at 0.40% and levels exceeding this can result in severe intoxication.

Ruminants fed diets lacking in sulphur can also show a marked reduction in growth rate, a decrease in milk production, their hair becoming rough and dull, excessive tearing and salivation, weakness and even death. As a consequence of this poor utilization of sulphur in the rumen, there is a decrease in the digestibility of proteins and carbohydrates.

The NRC (2007) reports that studies with goats revealed that the sulphur requirement for lactating goats should be greater than 0.16% and less than 0.36% of the DM. For kids, 0.22% of DM is recommended at an N:S ratio of 10.4:1. The sulphur requirement for 100 g/day, according to the committee, is 2.3 g/day and for adult goats it is 3.1 g/day. Therefore, the requirements for maintenance, growth and pregnancy were 0.22% of DM and 0.26% for lactating goats.

Costa et al. (2003), presents the sulphur requirements shown in tables 25 and 26. The author reports that the sulphur requirements for goats are not yet well defined, and also concludes that there is a difference in sulphur requirements for goats with one and two foetuses, always emphasizing the importance of carrying out more research with goats to reinforce the little data found so far.

Table 25. Estimated net daily requirement for sulphur (S) in pregnant goats with one and two foetuses.

Number of fetuses	Number of days	S (g)
	50	0,0116
	75	0,0227
1 Fetus	100	0,0468
	120	0,0875
	140	0,1706
	50	0,0170
2 Ferns	75	0,0480
	100	0,1057
120	0,1641	120
140	0,2105	140

Adapted from Costa (2003).

Table 26. Estimated daily dietary requirement for sulphur (S) in pregnant goats with one and two foetuses.

Number of fetuses	Number of days	S (g)
	50	0,0145
	75	0,0284
1 Fetus	100	0,0585
	120	0,1094
	140	0,2133
	50	0,0213
	75	0,0600
2 Ferns	100	0,1321
	120	0,2051
	140	0,2631

Adapted from Costa (2003).

Copper (Cu) requirements

According to Carvalho (2005), the Germans were the pioneers in discovering the presence of copper in plants and animal tissue, and it was first considered that the presence of this element in plants and tissues was the result of contamination by soil or water.

Khan (2007), studying the concentration of copper in improved pastures with seasonal variation in a semi-arid region, observed that the level of copper in the forage differed significantly between winter and summer, with copper concentrations in winter being considerably higher in the forage plants than in summer. The author also emphasizes that the levels of copper in the forage during the winter were within the range of the nutritional requirements of goats (Morand-Fehr, 1981).

Similar levels of copper in forage have also been found in various regions of the world, such as North Florida (Tiffany et al., 2001), Guatemala (Tejada et al., 1987), Central Florida (Espinoza et al. 1991) and Venezuela (Rojas et al., 1993). In Brazil, according to Carvaho (2005), copper concentrations are generally deficient in pastures.

Underwood (1981) reports that copper deficiency can also occur when there is a high intake of Mo and S, along with a normal intake of Cu.

McHargue (1925) reported on the essentiality of copper in 1920, however, evidence of the biological need for copper was effectively confirmed by Hart et al. (1928) working with rats, who in the same year discovered the presence of the element in the structure of hemoglobin.

In most species, the highest concentrations of copper are found in the liver, brain, heart, kidney, pigmented part of the eye, hair, pancreas and muscle and in low concentrations in the thyroid, pituitary, thymus, prostate, ovary and testicle.

Copper in plasma associates with around 90% of alpha 2-globulin as ceruloplasmin; only 10% is found in red blood cells as erythrocuprein.

Carvalho (2005) reports that in adult animals, the absorption of copper varies from 5 to 10% of that found in the diet, however, in young growing animals, this absorption can vary from 15 to 30%. The main site of absorption is in the small intestine, but the pH of the intestinal contents can modify absorption, as well as the interference of other minerals such as Ca salts, which decrease Cu absorption by raising the pH. The minerals Mercury, Molybdenum, Cadmium and Zinc also impair Cu absorption. Hg and Mo probably form insoluble compounds with copper; cadmium and zinc displace protein-bound copper in the intestinal mucosa. In view of this, it can be seen that copper absorption is directly linked to the chemical structural form in which the mineral is ingested.

Some combined forms of copper are better absorbed than others. Cupric nitrate, cupric chloride and cupric carbonate are better absorbed than cuprous oxide. Cupric sulphate is more easily absorbed than cuprous sulphate. Copper metal shows low absorption.

However, absorbed copper binds to plasma albumin to be transported to the liver, where it is stored. In the liver it is metabolized by the hepatic metalloprotein and enzyme ceruloplasmin, which is responsible for copper transport and is used in the synthesis of numerous proteins and enzymes, or released into the plasma as cupric albumin and in large quantities also as a component of ceruloplasmin, which is a copper-containing metalloenzyme linked to the metabolism of iron-ferroxidase. In the bone marrow, copper is taken up to form erythrocuprein in red blood cells, which has enzymatic action. This mineral is also involved in the nervous system with the maintenance of myelin in neurons (NRC 2007).

Copper is excreted in the feces via bile. Smaller amounts are lost through intestinal cells and pancreatic secretions, and insignificant amounts through urine.

It is not very common to find copper deficiencies in intensively reared animals, due to: the richness of copper in the raw materials; the minimal needs of the animal; the storage capacity in the liver.

Deficiency can be associated with the presence of phytates, excess protein in the feed or interactions with other minerals such as zinc, iron, molybdenum or sulphate.

The most specific signs of deficiency are: iron deficiency anemia, cardiovascular problems, gastrointestinal problems, loss of rigidity in the long bones, changes in skin pigmentation, discoloration and loss of hair, dermatitis and impairment of the quality of the hair, ataxia with fall of the posterior third due to demyelination of the nerves, reproductive failures with increased infertility and fetal mortality.

The NRC (2007) reports that the normal level of copper in goats is 0.9 to 1.39 mg/L. The committee reports that no data is yet provided to justify factorial methods for copper requirements in goats, thus recommending 15 mg/kg DM for lactating goats, 20 mg/kg DM for mature goats and 25 mg/kg DM for growing animals.

Underwood (1977) shows that for the heart, hair, brain and tissues copper levels are within a range of 9 to 15 ppm, and are also influenced by the rate of intake of this element in the diet. The idea is that copper concentrations tend to decrease with advancing age. In the work presented by Solaiman (2001), some of the values obtained differed from the recommendations given by Underwood (1977).

Solaiman (2001), studying the effect of medium and high copper supplementation in goats, can see the copper concentrations in some organs in Table 27.

The lack of studies on copper requirements for goats means that recommendations can only be made in the most appropriate way, which is why experiments have been used to provide different levels of copper and observe the animals' responses. One of the risks, for example, is what Suttle (1986) and Mills (1987) reported when they suggested that plasma Cu is a reference of limited value in diagnosing the state of copper in the animal organism, because an example such as inflammatory diseases could alter these levels. Even adequate plasma copper concentration can be misleading because other factors can contribute to the concentration found (McDowell and Arthington, 2005).

Table 27. Effect of copper supplementation on goat tissues

Fabric	Accumulated copper (g)		
Digestive tract	C (0.0 ppm)	SM (7.35ppm)	SA(14.70ppm)
Rumen	8,8±4,40	12,0±3,90	419,6±403
Reticle	2,4±0,05	7,0±0,90	356,6±346,8
Omaso	3,1±0,40	8,1±2,70	474,5±462,3
Abomasum	4,5±1,90	9,5±0,85	20,0±0,0
Duodenum	13,7±7,30	13,9±6,60	14,3±0,0
Jejunum	4,8±1,60	8,2±1,30	37,5±31,2

ileum	4,4±1,25	5,4±0,0	5,8±0,0
Colon	8,5±4,80	9,8±6,40	90,1±76,4
Internal organs			
Brain	8,3±0,10	7,7±0,6	7,5±0,8
Heart	4,2±0,40	6,0±0,8	14,4±9,5
Lungs	2,6±0,65	3,7±0,5	16,8±12,8
Liver	315,0±15,0	890±100	1100±100
Bacchus	3,2±0,80	1,6±0,35	17,3±13,8
Ovary	2,6±0,13	5,4±2,30	5,6±4,2
Single	4,6±1,50	4,9±2,4	8,1±4,30

Other fabrics and			
fluids			
Muscle	1,8±0,05	2,2±0,15	3,2±0,7
Fat	0,2±0,02	0,3±0,0	0,3±0,0
For	2,1±0,35	1,8±0,1	2,9±1,2
Rumen fluid	0,3±0,0	37,3±8,1	537,1±438,0
Abomasal contents	1,2±0,05	59,5±30	3805,0±2500
Feces	20,7±2,20	187,2±59,9	398,1±137,4
Bile	0,5±0,05	0,4±0,1	16,1±12,1
Blood	1,1±0,04	1,1±0,0	2,3±0,9

Adapted from Solaiman (2001).

Deficiency can be due to an excess of Mo, S, Fe, Zn and Ca, affecting all phases of growth, production and reproduction (Grahan, 1991). Also in relation to the immune system, copper deficiency can affect neutrophils and macrophages, resulting in a decrease in the number of antibody-producing cells.

The copper requirement for goats is estimated at an average of 8 to 10 mg / kg DM of the diet (Chew, 2000; Kessler, 1991; AFRC 1997). Young animals seem to be less sensitive to copper toxicity than sheep. This could be explained by a lower hepatic uptake of food with diets containing high levels of copper. Corroborating this, Zervas et al. (1989) reported values of 6-9 times lower for hepatic uptake in goats, which could lead to a greater susceptibility of newborn offspring to ataxia, especially in the case of multiple births.

Cobalt (Co) requirements

Cobalt (Co) belongs to the same family in the periodic table as nickel and iron. The first clear evidence that this mineral was essential in diets was observed in 1935 in an Australian study into the causes of certain bovine and ovine diseases known as "coastal disease" and "wasting disease" (Underwood, 1977). Cobalt is widely distributed in the animal organism, with the highest concentration in the liver, bone and kidneys (Underwood, 1977).

In ruminant nutrition, Co occupies a unique position among the mineral elements, as it is only used as an integral part of a vitamin (cyanocobalamin - Vitamin B12) by animals that depend on the symbiotic activity of their gastrointestinal microorganisms for this vitamin (Church, 1993).

However, cobalt is not highly absorbed by ruminants. The element is used by microorganisms to synthesize vitamin B12 in the rumen of ruminants. When cobalt is injected orally or in very large quantities, it tends to accumulate in the liver. Another peculiarity is that cobalt can replace zinc in certain proteolytic enzymes such as pancreatic carboxypeptidase Vallee (1974).

Cobalt is found in varying amounts in plants. However, most feeds contain adequate concentrations of cobalt, except in cases where forage crops are grown on cobalt-deficient soils. Supplementary sources include cobalt oxide and salts such as cobalt sulphate and cobalt chloride. Supplementary sources can be administered through incorporation into feed or mineral mixtures.

Its deficiency in ruminants causes vitamin B12 deficiency due to the inability of rumen microorganisms to synthesize sufficient amounts of this vitamin. There is no evidence that vitamin B12 is synthesized in the body's tissues. The importance of supplementing with this mineral comes from this synthesis, as studies have shown that when vitamin B12 is given orally, it is used less than when it is synthesized by rumen microorganisms (Church, 1993).

Vitamin B12 is necessary for the functioning of various enzyme systems in energy utilization. Ruminants deficient in vitamin B12 cannot efficiently convert propionate into succinate (methylmalonyl-CoA into succinyl-coA, a key step in neoglycogenesis in ruminants (Kozloski, 2002)). The main route of Co excretion and metabolic balance control is via urine.

Co deficiency occurs most frequently in ruminants under grazing, especially in large areas of tropical countries, and is found in soils of various origins, such as volcanic soils, sandy loams and washed sands.

By increasing the pH through liming, the amount of Co in the plant is reduced, which can aggravate the deficiency. Severe forms of Co deficiency in ruminants have been given various names, for example, in Brazil, the disease is commonly known as "peste de seca", "mal da areia", "mal das cabeceiras", and "mal do colete". The disease is characterized by loss of appetite, progressive weight loss, cachexia, lethargy, anaemia, rough hair growth, decreased milk production and low fertility. The disease occurs more frequently during the rainy season (Tokarnia and Dobereiner 1976).

When the deficiency is marginal, clinical signs may or may not occur and not only young animals, which are more susceptible, may show growth retardation which can be confused with the effects of parasites or low consumption. of the diet.

Co can be recycled via saliva or rumen epithelium (Church, 1993). Despite this, the best way to prevent a Co deficiency in grazing ruminants is through mineral supplementation containing at least 0.002% Co. Frequent injections of vitamin B12 can effectively prevent or cure Co deficiencies, but are very expensive.

There is little data on the effect of excessive cobalt feeding as a function of animal tissue levels. Keener et al. (1949) reported that oral administration of cobalt to cattle increased concentrations in the liver and kidneys by more than 10 times.

However, cobalt is not very toxic to cattle. Only doses of 4 to 10 mg of Co/kg of weight severely reduce the animal's appetite and weight, producing a state of anemia.

According to the NRC (2007), the cobalt requirement for goats is 0.11 mg Co/kg DM, which is different from the recommendations made by the previous committees, which was 0.1 to 0.15 mg Co/kg DM recommended by Haenlein (1992) and Meschy (2000), recommending 0.10 mg because of the sensitivity of goats to cobalt.

Chapter 5

Iron (Fe) requirements

Fe is considered to be one of the most abundant trace elements in the animal body and one of the two most abundant in nature. The body of an adult animal contains between 2.5 g and 4 g of this element, with around 2.0 g and 2.5 g (60 to 70%) circulating as a component of hemoglobin. Around 20% is stored in unstable forms in the liver, berry and other tissues. Approximately 300 mg are associated with electron transport and enzymes, associated with cytochromes and iron-sulphur proteins of electron transport and oxidative ionization in all cells, and enzymes of drug metabolism in the liver. The remaining 10% is in forms not available in tissues as a component of myosin and actinomyosin in muscles and associated with metalloenzymes (Church; Pond, 1982).

Iron is stored in the liver, spleen and bone marrow, forming complexes with proteins such as ferritin, a p-globulin, and as a component of hemosiderin, which results from the breakdown of ferritin. Hemosiderin contains 35% iron in the form of ferric hydroxide, is present in tissues as a brown pigment and is considered an insoluble form of Fe, while ferritin can be considered a soluble form of storage. Under normal and deficiency conditions, Fe is stored in equal quantities in both forms.

The iron compounds present in the body can be grouped into two categories: those that perform metabolic or enzymatic functions (hemoglobin, myoglobin and enzymes), and those associated with reserve iron. As a constituent of hemoglobin, iron is required for the transport of oxygen and carbon dioxide and is therefore directly involved in the process of cellular respiration (Guyton, 1991; Krause & Mahan, 1991).

Around 67% of the body's total iron is present in hemoglobin, which is made up of four subunits, each with an associated heme group. This molecule has only four iron atoms, which are considered essential because they combine with oxygen in the lungs and release it into the tissues (Anderson, 2005).

Enzymatic iron accounts for around 0.2% of total body iron and is found in the form of heme proteins containing a ferroporphyrin complex (such as cytochromes, which act in the transport of electrons in the respiratory chain), ferroflavoproteins and enzymes that require iron as a cofactor (Anderson, 2005).

Ferritin and hemosiderin are the reserve iron compounds and are involved in maintaining iron homeostasis in the body (Krause & Mahan, 1991). Iron also acts in metabolic processes such as the synthesis of purines (structural compounds of DNA and RNA), carnitine, collagen and neurotransmitters (dopanin, serotonin and norepinephrine) (Anderson, 2005).

Heme iron represents 5 to 10% of dietary iron, but its absorption can reach 25% compared to only 5% for non-heme iron (Pineda, 1994).

The absorption of heme iron occurs in the mucosa of intestinal cells as an intact iron-porphyrin complex, and is little affected by the composition of food and gastrointestinal secretions. In the cytosol, ferrous iron is enzymatically removed from heme iron and combines with apoferritin to form ferritin. This protein acts as an intracellular store of iron and transports it to the basolateral membrane, where by active transport the iron is moved into the blood.

Unlike heme iron, non-heme iron must first be ionized by gastric secretion into the ferrous or ferric form in order to be absorbed. When it passes into the duodenum, where the pH is close to neutral, most of the ferric form is precipitated while the ferrous iron, more soluble at pH 7, remains available for absorption. Non-heme (ionic) iron enters the enterocyte mucosa by facilitated diffusion, directed inwards by a concentration gradient, and follows the same transport and passage route to the blood as that described for heme iron (Anderson, 2005).

Ascorbic acid is the most effective promoter of iron absorption, and it has recently been shown that it can also influence the transport and storage of iron in the body (Lynch, 1997). The absorption of non-heme iron is also increased by the presence of vitamin A, citric, malic, tartaric, lactic and other organic acids (Guyton, 1991). On the other hand, substances such as tannins, polyphenols, phytates and dietary fibers have been reported to inhibit iron absorption (Hallberg et al., 1989).

After absorption, iron enters the bloodstream bound to a carrier protein called transferrin, which is able to bind two iron atoms via its receptors. These receptors deliver iron to the bone marrow to be incorporated into new hemoglobin molecules, to other tissues that require iron and to storage sites. After release, transferrin becomes available again to transport iron (Aires, 1991).

The amount of free transferrin in the blood seems to regulate the absorption of iron from the intestinal membrane. Transferrin is generally saturated with iron up to a third of its total capacity. In individuals with normal iron levels in the body, transferrin remains saturated, with less absorption and greater loss of iron from the mucous cells by desquamation. In those with iron deficiency, transferrin is less saturated and more iron is absorbed from the intestinal mucosa. The opposite phenomenon is seen in people with excess iron, i.e. excretion increases and absorption is limited, demonstrating the existence of a self-regulating mechanism for iron absorption (Krause & Mahan, 1991; Anderson, 2005).

Iron homeostasis in the body is maintained by the absorption process, as only small amounts are excreted. Excess iron is stored in the form of ferritin and is easily mobilized when the body's needs for this mineral increase. Smaller amounts of iron are stored as hemosiderin, a more stable and less accessible form of deposit. Iron is lost from the body through bleeding and, in very small quantities, through fecal excretion, perspiration and normal exfoliation of hair and skin (Aires, 1991; Anderson, 2005).

In conditions of excess iron, hemosiderin predominates (Church; Pond, 1982; Linder, 1991). Absorbed iron is tenaciously retained. Fecal iron results from unabsorbed iron, although a small proportion (0.3 to 0.5 mg/kg) is lost through bile and epithelial cell desquamation. However, when iron is injected, little is excreted via the fecal and urinary routes, and losses only occur when iron is supplied parenterally in excess of the Fe-binding capacity of the plasma, or when chelating agents are supplied (Church, 1982).

The body hardly excretes Fe, but according to Carvalho et al. (2005), iron can be found in feces (unabsorbed Fe), urine (kidney problems) and sweat. However, animals normally have a limited capacity to excrete Fe and its absorption is very poor, being basically controlled by homeostasis. Absorption takes place in the duodenum and jejunum, and depends on various factors such as the age of the animal, where the younger the animal, the greater the absorption capacity; the level of Fe in the individual where, when deficient, there is an increase in absorption; the health of the animal, where weakened animals have a lower capacity; the condition of the intestine, where any process that affects the integrity and functioning of the mucosa impairs absorption; the quantity and chemical form of Fe; and the presence of other elements with which Fe can interfere with absorption.

The main site of Fe excretion is the uterus, which supplies it to the fetus, or loses it in the metastrous hemorrhage as the animal cycles. All Fe is permanently recycled (Carvalho et al., 2005).

According to the NRC (2007), the requirements for goats were estimated by the AFRC (1980) through work carried out by Lamand (1981) and ARC (1980) with a value of 30 mg/kg of DM in the diet, emphasizing that Haenlein (1987) justifies the high requirement for goats because these animals are more frequently affected by internal parasites, suggesting higher values ranging from 30 to 100 mg/kg DM of the diet, depending on the age and iron status of the animal.

The recommendations for iron requirements established by the NRC (2007) are:
Growth = 95 mg/kg in the feed
Pregnancy = 35 mg/kg in the feed
Lactation = 35 mg/kg in the feed

Costa (2005), estimating iron retention in goats during pregnancy using an exponential model, found the values shown in table 28.
He concluded that the estimates made by the proposed model efficiently represented the biological behavior of mineral retention in goats, also noting the differences in iron requirements between goats carrying one and two fetuses.

The same author in 2003 estimated the iron requirements of goats with one and two fetuses with the results shown in tables 29 and 30.

Table 28. Comparison between sodium retention in the pregnant uterus + udder (Y) and its estimation using the regression model $\ln Y = A + Bx + Cx^2$ (Y1) in goats at different stages of pregnancy, with one and two foetuses.

Length of pregnancy/No. of fetuses	Estimate	Fe (mg)
50 days/1 fetus	Y Y1	13,46 16,89
100 days/1 fetus	Y Y1	134,29 130,69
140 days/1 fetus	Y Y1	575,19 575,50
50 days/2 fetuses	Y Y1	15,36 15,57
100 days/2 fetuses	Y Y1	174,73 174,75
140 days/2 fetuses	Y Y1	981,10 981,00

Adapted from Costa (2005).

Table 29. Estimated net daily iron (Fe) requirement in pregnant goats with one and two fetuses.

Number of fetuses	Number of days	Fe (mg)
	50	0,7798
	75	1,8293

	100	4,5838
1 Fetus		
	120	10,0332
	140	22,9443
	50	0,8096
	75	2,6797
2 Ferns	100	8,1522
	120	18,6760
	140	40,5065

Adapted from Costa (2003).

Table 30. Estimated daily dietary requirement for iron (Fe) in pregnant goats with one and two fetuses.

Number of fetuses	Number of days	Fe (mg)
	50	1,5596
	75	3,6586
1 Fetus	100	9,1676

	120	20,0664
	140	45,8886
	50	1,6192
2 Ferns	75	5,3594
	100	16,3104
	120	37,3520
	140	81,0124

Adapted from Costa (2003).

Manganese (Mn) requirements

According to Underwood (1977), the concentration of manganese in forage can vary greatly depending on the species, variety, soil type and concentration of the element in the soil. However, the chemical form in which manganese is found and the way in which it is associated with the structural components of plants influences its availability to goats, as well as its participation in metabolic processes.

This mineral plays an important role in the oxidation of water molecules in plants. It is most often found in free form (Mn^{++}) or as low molecular mass, positively charged complexes (Tiffin, 1972; Tinker, 1981). It is normally transported to meristematic tissues, and its concentration is greatest in young growing tissues. Manganese appears to be poorly transported by the phloem, which may explain its relatively low concentration in fruits, seeds and root reserve organs. Manganese mobility is lower when there is little availability of the element to plants, reducing manganese translocation in the plant (Kabata-Pendias & Pendias, 1984). However, Cheniae & Martin (1968) report that studies on manganese in plants are focused on its participation in the unfolding of the water molecule and in the evolution of O_2 in the photosynthetic system, in order to elucidate the physiological factors in which manganese is involved so as to know where and how much of the mineral is present in forage plants, in order to serve as a source of this mineral for animals.

In order to elucidate these issues, Heenan & Campbell (1980) reported that with increasing age there is a greater accumulation of manganese in the leaves, however, a small amount of the element is translocated from the older leaves to the younger ones, where the element is normally deficient. However, the concentration of manganese varies greatly within the plant and during the growing season, leading to variations in the availability of manganese in forage plants and making it difficult to supply this nutrient to meet the nutritional requirements of goats.

However, in animals, manganese is distributed in very low concentrations in the body's cells and tissues and is necessary to maintain the perfect functioning of the productive processes of both males and females.

Absorption occurs throughout the small intestine, but is quite low (< 1% in juveniles and slightly higher in females). Absorption and excretion are controlled by homeostasis, with no sudden variations in Mn levels (Church, 1993) and its excretion is via bile. The consumption of diets rich in Ca and Fe reduces the absorption of Mn. The animal body has a limited capacity to store mobilizable reserves of Mn. Bones, liver and kidneys normally have higher concentrations of Mn than blood and muscles (Underwood, 1977; Hidiroglou, 1979).

Manganese deficiencies can result in a wide variety of metabolic and structural alterations. These general clinical signs of deficiency are: skeletal abnormalities, reproductive disorders associated with manganese deficiency include abnormalities in the estrous cycle, silent estrus, low conception rates, an increase in the number of abortions and delayed sexual maturity (Hurley and Doane, 1989), deterioration of the central nervous system and slow growth (Wilson, 1966; Anke et al., 1973). It inhibits the synthesis of cholesterol and its precursors and these, in turn, limit the synthesis of sex hormones, causing infertility.

Table 31 shows an example of the effect of manganese deficiency on goats.

Table 31. Influence of manganese in goat supplementation

	Normal diet	Mn deficiency in the diet
Colostrum (Mn in DM) ppm	1,13	0,63
Milk (Mn in DM) ppm	1,03	0,86
Pelo (Mn in DM) ppm	1,70	0,96
Carcass (Mn in DM) ppm	2,5	1,8

Adapted from Schellner (1972) quoted by Haenlein (1980).

Only in areas where a Mn deficiency or conditioned deficiency is suspected should Mn be added in the form of sulphates, carbonates, oxides or chlorides to the mineral mixture. Fertilizing pastures is also an advisable practice (Pott, 1986).

Excess manganese affects the metabolism of some mineral elements. The first important antagonism in manganese toxicity is the effect on Fe, reducing the level of hemoglobin and causing a hepatic accumulation of Cu. High levels of manganese in the diet have an adverse effect on growth, producing anemia and altering the rumen microflora.

Underwood (1971, 1977), in a review of manganese concentrations in animal tissues, observed that the liver and pituitary gland contain the highest concentrations, each at around 2.5 ppm. In the coat, he observed that the presence of manganese is influenced by dietary levels of deficiency or excess in a more sensitive way than the internal organs or tissues. Efficient homeostatic mechanisms for eliminating excess manganese from the body seem to prevent significant accumulation in animal tissues (Watson et al., 1973).

Table 32 shows the concentration of manganese in the organs of goats fed diets deficient in this mineral.

Table 32. Influence of manganese supplementation on organ concentration in goats.

Organs	Normal diet	Mn-deficient diet
Liver	11,5	3,7
Kidney	4,4	2,6
Berry	3,8	2,3
Courage	2,1	1,1
Muscle	1,7	1,6
Brain	3,8	3,0
Spinal cord	2,8	1,5
Ovaries	4,3	2,7
Testicles	2,9	2,9
Pancreas	5,8	5,4

Aorta	2,2	1,8	
For	2,9	1,5	

Adapted from Anke (1073) quoted by Haenlein (1980).

The NRC (2007) presents the equations for manganese requirements for goats shown in table 33.

Table 33. Manganese requirements for goats according to NRC recommendations (2007)

Species and Class	Requirements	Equation
Mantenga	0.002 mg/kg	PC/0.0075
Growth	0.7 mg/kg	GPV/0.0075
Gestation	0.025 mg/kg	GPV/0.0075
Lactagao	0.03 mg/kg	PL/0.0075

Adapted from NRC (2007).

Selenium (Se) requirements

According to Carvalho (2005), selenium was discovered in Sweden in 1818 by the chemist J.J. Berzelius. However, the first evidence that selenium was an essential nutrient was provided by Schwarz and Foltz, (1957) when they discovered that it would prevent liver necrosis in rats, and was subsequently used successfully by Muth et al. (1958) and Hogue (1958) to prevent muscle disease in young ruminants.

Selenium is essential for various functions in the body, such as growth, reproduction, disease prevention and muscle integrity protection (McDowell, 1992; Boland, 2003). It is a component of the enzyme glutathione peroxidase, which is involved in protecting cell membranes against oxidative degeneration (Dayrell, cited by Peixoto, 1985). When there is too much or too little, intoxication occurs and deficiencies can arise.

Carvalho et al. (2005) report that the Selenium ingested, whether in plants, grains or via mineral supplements, is greatly interfered with in rumen fermentation. Around 65 to 70% of the Selenium ingested is lost in the rumen, probably due to oxidative processes that reduce Selenium to insoluble compounds, so that only around 30 to 35% is biochemically able to be absorbed in the small intestine of ruminants, and the rest is lost in the feces.

Selenium is absorbed mainly between the duodenum and ileum. Its availability is related to its chemical form: the more reduced it is, the less available it is to animals. If the Selenium source is inorganic, of the selenite type, diffusion is facilitated by the wall of the small intestine, mainly by the duodenum, if the Selenium source is inorganic of the selenate type, absorption is by active transport and requires energy and if the Selenium source is organic in the form of selenomethionine, the problem is that this absorption route is the same as that of the amino acid methionine, which will compete for transport and may inhibit the absorption of selenomethionine.

In addition to other complications, selenium deficiency can cause reproductive problems in both males and females (Rosa, 1993; Luberda, 2005). Kamada and Kodate (1998) found a low level of plasma progesterone, which reduced conception, increased embryo mortality and abortions, increased placenta retention and caused disturbances in the estrous cycle.

VanNiekerk et al. (1996) found a 20 to 40% reduction in embryonic mortality in sheep with a nutritional deficiency of selenium. Grace (1994) found that selenium deficiency has a negative effect on animals, as it is responsible for the formation of the enzyme 5' deiodinase, which transforms T4 into T3 (Arthur, 1993), which impairs animal development (Oblitas et al., 2000). Smith and Akinbamijo (2000) found that goats are also susceptible to selenium deficiency, which functions as a cytosolic component to form GSH-Px, which reduces peroxides, preventing them from attacking cell membranes.

However, nutritional muscular dystrophy, called white muscle disease, which occurs in small and large ruminants, is the main symptom of Se deficiency. It is characterized by the degeneration of skeletal muscle, especially those that are most active. It usually affects animals shortly after birth, with difficulties in moving around and even death. Bilateral symmetrical lesions of a whitish color can be seen in the skeletal muscle, hence the name.

Some authors report that a dietary intake of 0.1 to 0.2 ppm Selenium can provide an adequate safety margin against dietary variations for grazing animals. The minimum requirement for Selenium varies according to the form in which it is ingested and other dietary factors. It is also known that high consumption of sulphates reduces the availability of Se for animals, so that when there is a high concentration of sulphate in the diet, Se requirements are higher.

According to the NRC (1983), the recommendations are 0.1 to 0.3 ppm of selenium in the diet, however, Mestchy (2000) highlighted the daily requirement of 0.1 mg of Se/kg of DM in the diet of pregnant goats.

In a study carried out with 12 healthy dairy goats over 10 weeks, in which the control group received a diet containing 0.045 mg of Sel-Plex (Selenomethionine)/kg of dry matter (DM) and a second group received a diet supplemented with 0.15 mg of Sel-Plex/kg of DM (Khaled and Illek., 1999), they found that the animals with the highest supplementation had higher levels of Se in their plasma and milk, but that this did not affect thyroid activity.

The recommendations made by the NRC (2007) are shown in table 34.

Table 34. Selenium requirements for goats according to NRC recommendations (2007)

Species and Class	Requirements	Equation
Mantenga	0.015mg/kg	IMS+0.083mg/CA*

Growth	0.5mg/kg	GPV/CA*
Gestation	0.0021mg/kg	GPV/CA*
Lactagao	0.10mg/kg	PL/CA*

Where: *CA = absorption coefficient (forage = 0.31 and concentrate = 0.60)

Iodine (I) requirements

Probably the first person to recommend the use of iodine in salt as a means of preventing goitre was Koestl, who began using it in Austria in 1895.

The use of iodine in animal production is not limited to its role as a nutrient in food. According to Miller and Tillapaugh (1966) iodine can also be used to prevent or treat various diseases that affect animals.

Although it is found in practically every part of the body, its essentiality lies in the synthesis of thyroid hormones, where it is the main storage site for this mineral, which participates in the regulation of energy metabolism. Thyroid hormones, which contain iodine, are known to play a role in thermoregulation, intermediary metabolism, reproduction, growth and development, hematopoiesis, circulation and neuromuscular function.

According to Underwood (1977), iodine in food occurs largely as inorganic iodide and is absorbed at all levels of the gastrointestinal tract. However, in ruminants, the rumen is the main site of iodine absorption with around 70 to 80% and 10% in the abomasum (Barua et al., 1964). Once absorbed, iodide is rapidly distributed throughout the body, with the main sites of iodine concentration being the thyroid and kidneys. In addition, iodine is also concentrated in the salivary glands, stomach, skin and hair, mammary glands, placenta and ovary (Gross, 1962).

The main clinical sign of iodine deficiency is hypothyroidism or goiter. However, for clinical signs of an enlarged thyroid gland to be detectable, the animals must be exposed to iodine-deficient diets for a long period of time (Miller et al., 1993b).

Although signs of goiter only appear after months of feeding iodine-deficient diets, in cases of subclinical deficiency, where goiter is not yet detected, reproductive performance can be compromised (Miller et al., 1993b; Corah and Ives, 1991; Ferguson, 1991). According to these authors, females fed diets marginal in iodine can show signs of infertility, an increase in the rate of early embryonic mortality, abortions, the birth of weak calves or even stillbirths, an increase in the incidence of placental loss, a reduction in the conception rate, uterine inertia and delayed fetal development.

Widely ingested in the form of iodide, iodine is rapidly transported to the thyroid gland by plasma proteins, where it is oxidized by thyroid peroxidase to iodine, combining with tyrosine to form inorganic iodine: monoiodotyrosine, diiodotyrosine, triiodothyronine (T3) and tetraiodothyroxine (T4; thyroxine) (Jones et al, 2000; Church, 1993; Carvalho et al., 2005).

However, it can be said that the main function of iodine, through thyroid hormones, is to control the rate of cellular oxidation of all the body's tissues. In animals with hypothyroidism, there is a reduction in the degree of energy exchange and in the amount of heat released by the tissues, and the rate of basal metabolism declines.

According to the NRC (1985), the recommendations for iodine were 0.10 to 0.80 mg I/kgMS. However, despite the new edition, the NRC (2007) reports that there is still not enough data to guarantee optimal iodine levels, presenting the mineral requirements for goats shown in Table 35.

Table 35. Iodine requirements according to NRC recommendations (2207)

Species and Class	mg/kg diet
Growth and maturity and non-lactating	0,5
Lactation	0,8
Adapted from NRC (2007).	

Zinc (Zn) requirements

Zinc is a metal considered to be one of the essential trace elements in the body. It has an atomic number of 30, an atomic mass of 65.39, an atomic radius of 133.2 pm and a specific mass of 7.133 g/cm3. In biological systems, zinc is found in the nucleus of cells, chromosomes, ribosomes and secreted granules (Costa, 2004).

It is usually found in association with proteins or other organic compounds. Its importance as a supplement in animal feed was demonstrated in 1955 by Tucker and Salmon.

Absorption by the animal is directly related to its needs. A growing or highly productive animal absorbs more zinc than one with lower requirements. Adult animals do not store zinc in their organs, so they are unable to meet their needs through reserves. As a result, when animals are fed diets low in zinc, deficiency symptoms are not long in coming.

Zinc is absorbed by the intestinal tract, part in the abomasum and the rest in the duodenum, in relation to its need, and is subsequently metabolized in the liver, which is the largest zinc metabolism organ (Conrad, et al., 1985). The main route of excretion is through the feces with small amounts of fecal zinc derived from bile, pancreatic secretions, desquamated epithelial cells, and also lost in small amounts in the urine and through sweating.

However, Zinc is responsible for activating some enzymes and is a component of a large number of important metalloenzymes (Riordan and Vallée, 1976). The latter include carbonic anhydrase, carboxypeptidases A and B, alcohol dehydrogenase, glutamic dehydrogenase, D-glyceraldehyde-3-phosphate dehydrogenase, lactate dehydrogenase, malate dehydrogenase, alkaline phosphatase, aldolase, superoxide dismutase, ribonuclease, DNA polymerase, and others. It is also associated with FSH and LH augmentation (effects potentiated by zinc), vitamin A uliliation, sulphate metabolism and brain development (Peixoto et al., 1995).

Several studies have reported the benefits of zinc. Yamaguchi et al. (1995) studied the behavior of zinc by evaluating the in vitro and in vivo behavior of this element. These authors concluded that zinc increases the amount of bone proteins, calcium content and alkaline phosphatase activity. Using a zinc concentration of IOOLIM increased the amount of bone proteins, alkaline phosphatase level and calcium content (Bettger, W.J; 1993; Yamaguchi, M; 1988).

However, in the case of zinc deficiency, the utilization of amino acids in protein synthesis is incomplete. For this reason, its deficiency has a particular impact on growth and reproduction.

However, the first symptoms of Zn deficiency include reduced consumption of the diet, stunted growth, as Zn also participates in various ways in cell proliferation, in the regulation of DNA synthesis, as well as influencing the hormonal regulation of cell division (Macdonald, 2000). It also causes poor food conversion efficiency, followed by skin disorders. Clinical signs of severe deficiency include alopecia, dryness and scaling of the skin on the head, ventral part, scrotum and legs (parakeratosis). Other clinical symptoms include inflammation of the nose and mouth, hardening of swellings and joints, increased growth of bacteria on the oral mucosa with gingivitis. Zn deficiency also leads to a decrease in the body's immune capacity.

Zn supplementation in mineral salt is the most suitable form for grazing animals, especially in tropical climate regions.

However, according to Mcdowell et al. (1983), the tolerance limit depends mainly on the relative content of Ca, Cu, Fe and Cd, with which zinc competes in the utilization and absorption process.

The most sensitive responses to excess dietary zinc were increases in zinc concentrations in the serum, liver, kidney, pancreas and small intestine, with bone almost the only response. Zinc was sometimes higher in the heart, but always remained unchanged in skeletal muscle.

According to the NRC (2007), zinc requirements for growth in goats are different from those found by the NRC (1981) which presents requirements of 10 mg Zn/kg DM in the diet for growth, production and reproduction and by the NRC (1985) which reports values of 20 mg Zn/kg DM in the diet for growing animals and 33 mg ZN/kg DM in the diet for reproduction.

The zinc requirements established by the NRC (2007) are shown in table 36.

Table 36. Selenium requirement for goats according to NRC recommendations (2007)

Species and Class	Requirements	Equation
Maintenance	0.045 mg/kg	PC/CA*
Growth	0.025 mg/kg	GPV/CA*
Pregnancy	0.05 mg/kg	GPV/CA*
Lactation	5.5 mg/kg	PL/CA*

Where: *CA = absorption coefficient (0.15 for adult goats)

Molybdenum (Mo) requirements

Mo is a mineral microelement found in almost all cells and body fluids and is considered essential for its biochemical role as part of the structural component of the enzyme sulfide oxidase and as a cofactor for the integral activity of the enzymes xanthine oxidase and aldehyde oxidase (Renzo et al., 1953; Richert and Westerfield, 1953). Xanthine oxidase is present in microorganisms, animal tissues and milk. The highest concentration of molybdenum in the animal organism is found in skeletal muscles, followed by the liver, kidneys, skin, hair and wool.

Other authors also emphasize the essentiality of molybdenum, always affirming its importance in establishing the production of xanthine oxidase at normal levels. All these considerations support the use of molybdenum in goat diets, but the need for molybdenum to maintain the body's metabolic homeostasis has not yet been strongly confirmed, since Bass at al. (1950) reported that even if part of the xanthine oxidase is removed, no changes are observed in the excreted concentrations of uric acid and allantoin.

Molybdenum is absorbed in the abomasum and duodenum and according to Carvalho (2005), molybdenum is absorbed rapidly, accounting for around 20 to 30% of the mineral present in the diet. In tissues, blood and milk, the concentration of molybdenum can be altered by the concentration in the diet (Dick, 1953[a]). However, very little of this mineral is stored in the body. Excretion is preferably via urine, with a small amount being excreted via bile and milk (Churchu, 1993).

Gray and Daniel (1964) reported that molybdenum concentrations in the liver of animals on normal diets can vary from 2 to 4 ppm on a dry matter basis, but levels as high as 30 ppm can be found if the animals consume high levels of the mineral. Concentrations in the kidneys can approach 50 percent of those found in the liver (Underwood, 1977).

According to the MRC (2007), there is a significant variation in molybdenum requirements between species. It also reports that the other committees make no mention when it comes to molybdenum requirements for goats, emphasizing that within the literature for sheep, there is no report on dietary molybdenum requirements. However, a requirement of 0.5 mg Mo/kg of dry matter in the diet has been established according to (NRC, 1975; NRC 1985). The AFRC (1997) also did not define molybdenum requirements. Kessler (1991) and Hanlein (1992) defined a requirement of 0.1 mg / kg DM diet for goats, adding that goats are more susceptible to molybdenum deficiency. Young goats fed semi-purified diets at 1 mg/kg DM improved consumption and growth, while diets containing 0.06 mg / kg DM did not provide an improvement in consumption and growth.

Final considerations

In light of the above, it can be seen that several factors can affect mineral requirements in goats, since the effects of environmental factors and the effects of mineral elements on nutritional requirements do not seem to be very well understood due to the diversity and range of variables involved, leading to limitations. In addition, there is still limited knowledge about the availability of minerals, especially for goats raised on pasture

However, research focusing on mineral requirements in goats should be encouraged and conducted, given the importance of minerals in all animal metabolism, directly interfering in production and reproduction. It is also necessary to intensify the work carried out in Brazil with the aim of acquiring a more accurate and precise knowledge of mineral requirements within Brazilian peculiarities, thus reducing the inaccuracies that permeate mineral requirements for the goat species in the country. In this way, the studies have aimed to form and organize a table of requirements that can make feeding more feasible so that there is no underestimation or overestimation of mineral requirements.

Bibliographical references

Agricultural and Food Research Council - AFRC. **Technical Committee on Responses to Nutrients, Report Number 6. A Reappraisal of the Calcium and Phosphorous Requirements of Sheep and Cattle.** Nutr. Abstr. Rev. (Series B), v.61, n.9, p.576-612, 1991.

Agricultural and Food Research Council - AFRC. The nutrition of goats. **Nutrition Abstracts and Reviews (Series B)**, v.67, n.11, p.765-830,1997.

Agricultural Research Council. ARC. **The nutrient reguirements of farm livestock.** London, 1980. 351 p.

Agricultural Research Council. NRC. **The Role of Chromium in Animal Nutrition.** National Academy Press, Washington, DC. 1997.

Aires, MM. **Fisiología.** 5.ed. Sao Paulo: Guanabara Koogan; 1991. 800p.

Albiston, H.E. Toxaemic (enzootic) jaundice of sheep. In "Diseases of Domestic Animals in Australia". Australian Department of Health, Service Publication (Animal Quarantine) Number 12. **Australian Government Publishing Service**, Canberra. 1975.

Alcalde, C.R.; Ezequiel, M.B.; Lema, A.C.F. et al. Endogenous Losses and Apparent and Real Absorption Coefficients of Magnesium in Goats. **Rev. bras. zootec.**, v.28, n.6, p.1347-1357, 1999.

Amado, T.J.C.; Mielniczuk, J. & Aita, C. Recommended nitrogen fertilization for maize in RS and SC adapted to the use of soil cover crops, under a no-till system. R. Bras. Ci. Solo, 26:241-248, 2002.

Ammerman, C. B., J. M. Wing, B. G. Dunavant, W. K. Robertson, J. P. Feaster, and L. R. Arrington. 1967. Utilization of inorganic iron by ruminants as influenced by form of iron and iron status of the animal. **J. Anim. Sci.** 26:404.

Anderson JB. Minerals. In: Mahan LK, Escott-Stump S, Krause MV. **Food, nutrition and diet therapy.** São Paulo: Roca; p.107-45, 2005.

Andriguetto, J. M. **Nutridio animal.** 4. ed. Säo Paulo: Nobel, 1990. 395 p.

Anke, U.; Grppel, B.; Reisseg, W. et al. Mangamangel beim wiederkauer, Skelett-und Nervenstorungen bei Weiblichen Wiederkäuer und ihren Nachkomen. **Achiv Fur Tierernährung**, v.23, p. 197 - 211, 1973.

Arthur, J. R. The biochemical functions of selenium relationships to thyroid metabolism and antioxidant systems. In: Rowett Research Institute Annual Report, p. 11-20. **Rowett Research Institute**, Backburn, Abedeen, UK, 1993.

Barcellos, J. et al. **Mineral nutrition in ruminants.** Porto Alegre: Gráfica da UFRGS, 1998, 146p.

Barcellos, J.O.J. et al. Mineral supplementation of beef cattle in subtropical environments. In: BARCELLOS, J.O.J. et al. **Mineral supplementation of beef cattle in subtropical regions.** Porto Alegre: UFRGS, p.19-52, 2003.

Barcellos, J.O.J. The role of phosphorus in beef cattle nutrition. In: Diaz Gonzalez, F.H.; Ospina, H.; Barcellos, J.O.J. (Ed) **Nutrigao mineral em ruminantes.** 2 ed. Porto Alegre: Editora da UFRGS, 1998, p. 23-72.

Barua, J., R. G. Gragle, and J. K. Miller. Sites of gastrointestinal-blood passage of iodide and thyroxine in young cattle. J. Dairy Sci. 47:539, 1964.

Blair-West, J. R. et al. Physiological, morphological and behavioral adaptation to a sodium deficient environment by wild native. **Australian and introduced species of animals.** Nature 217:922-928, 1968.

Bohl, C.H; Volpe, S.L. Magnesium and exercise. **Crit Rev Food Sci Nutr**, v. 6, n. 42, p.533-63, 2002.

Boland, M. P. Trace minerals in production and reproduction in dairy cows. **Adv. In D. Techn.**, Dublin, Ireland, v.15, p.319-330. 2003.

Bronner, F. Calcium absorption: a paradigm for mineral absorption. *J. Nutr.*, v.128, n.5, p.917-920, 1998.

Bueno, M.S.; Vitti, D.M.S. Phosphorus Levels for Goats: Fecal Endogenous Loss and Net Requirement for Maintenance. **Pesq. agropec. bras.**, Brasilia, v.34, n.4, p.675-681, Apr. 1999.

Bueno, M.S.; Vitti, D.M.S.S. Phosphorus Levels for Goats: Fecal Endogenous Loss Net Requirement for Maintenance. **Pesquisa Agrapecuária Brasileira**, Brasília. V.34, N.4, P. 675-681, 1999.

Capen, C.C. The calcium regulating hormones: parathyroid hormone, calcitonin and cholecalciferol. In: Mcdonald, L.E. (Ed.) **Veterinary endocrinology and reproduction**.3.ed. Philadelphia: Lea & Febiger, p.60-130, 1980.

Cardoso, M.A. **Nutrição Humana: Nutrition and Metabolism**. Series II. Rio de Janeiro: Ed Guanabara Koogan S.A, Chapter 15, p. 237, 2006.

Carvalho, F.A.N.; Barbosa, F.A.; Mcdowell, L.R. **Nutrigao de Bovinos a Pasto.** 1ª edigao, Belo Horizonte: PapelForm, p. 428. 2003.

Carvalho, F.A.N.; et al. **Nutrigao de bovinos a pasture**. 2° ed. Belo Horizonte: PapelForm. 2005. 428p.

Carvalho, F.F.R.; et al. Endogenous Loss and Phosphorus Requirement for the Maintenance of Saanen Goats. **Rev. bras. zootec.**, v.32, n.2, p.411-417, 2003.

Cavalheiro, A.C.L.; Trindade, D.S.; Becker, A.S.; et al. Effect of mineral supplementation on the productive performance of grazing steers. **Rev. bras. zootec.**, v.21, n.3, p.456466, 1992.

Chew, B.P., Micronutrients play role in stress, production in dairy cattle. Feedstuffs, June 12. p. 11.,2000.

Chicco, C. F., C. B. Ammerman, W. G. Hillis, and L. R. Arrington. Utilization of dietary magnesium by sheep. Am. J. Physiol. 222:1469, 1972.

Church, D.C. **Fisiologia digestiva y nutrición de los rumiantes**, 4.ed. Zaragoza: Acribia, p.127-135,1993,

Church, D.C.; Pond, W.G. **Basic animal nutrition and feeding**. 3.ed. New York: John Wiley & Sons, 1982. 403p.

Coelho Da Silva, J.F. et al.Effect of sodium monensin and urea on consumption, rumen parameters, apparent digestibility and nutritional balance of cattle. **Rev. Bras. Zoot.**, v.20, p.454-462, 1991.

Comar, C.L., Monroe, R.A., Visek, W.J. et al. Comparison of two isotope methods for determination of endogenous fecal calcium. *J. Nutr.*, v.50, p.459-67, 1953.

Conrad, J.H. Feeding of farm animals in hot and cold environments. In: Yousef, M.K. (Ed.), **Stress Physiology in Livestock**. CRC Press, Inc., Boca Raton, Florida, U.S.A., Place, Corvallis, OR 97330, USA. 1985.

Corah, L.R., Ives S. The effects of essential trace minerals on reproduction of beef cattle. In: The Veterinary Clinics of North America. **Fo. An. Pract.- Beef Catt. Nutr.**, vol. 7, p. 41-57, 1991.

Costa, R.C.; et al. Mineral requirements for goats during pregnancy: Na, K, Mg, S, Fe and Zn. **Rev. bras. zootec.**, v.32, n.2, p.431-436, 2003.

Costa, R.G.; et al. Mineralization for sheep and goats. In : IX Simpósio Nordestino de Alimentação de Ruminantes, Campina Grande, 2004. **Proceedings...** Campina Grande : SNPA, 2004. CD-Rom.

Costa, R.G.; et al. Mineralization for sheep and goats. In : IX Simpósio Nordestino de Alimentação de Ruminantes, Campina Grande, 2004. **Proceedings...** Campina Grande : SNPA, 2004. CD-Rom.

Costa, V. M. M. **Eimeridiidae in goats and sheep**. 43f. 2006. Course Conclusion Work (Graduation in Veterinary Medicine) - Department of Animal Science, Universidade Federal Rural do Semi Árido, Mossoró-RN, 2006.

Costa,R.G.; Resende,K.T.; Rodrigues, M.T. et al. Mineral Retention by Goats During Gestation. **Agropecuária Técnica**, v.26, n.2, p.129-133, 2005.

Dayrell, M.S. **Mineral content in animal tissues, plants and soils in Brazil**. EMBRAPA-CNPGL. Document 24. Coronel Pacheco, 1986.

Dell'isola, A.T.P.; Veloso, J.A.F.; Baiao, N.C. et al. Effect of soybean oil in diets with different calcium levels on absorption and bone retention of calcium and phosphorus in broilers. **Arquivo Brasileiro de Medicina Veterinária e Zootecnia**, v.55, n.4, p.461-466, 2003.

DeRenzo, E. C., E. Kaleita, P. G. Heytler, J. J. Oleson, B. L. Hutchings, and J. H. Williams. Identification of the xanthine oxidase factor as molybdenum. **Arch. Biochem. Biophys.** 45:247, 1953.

Dias, H.L.C. **Nutritional value of tropical pastures.** Seminar for ZOO 650: Forage: UFV. 15p, 1997.

Dick, A. T. The effect of inorganic sulphate on the excretion of molybdenum in sheep. **Aust. Vet. J.** 29:18, 1953a.

Duncan, C. W., C. F. Huffman, and C. S. Robinson. Magnesium studies in calves. I. Tetany produced by a ration of milk or milk with various supplements. J. Biol. Chem. 108:35, 1935.

Espinoza, J.E., L.R. McDowell, N.S. Wilkinson, J.H. Conrad and F.G. Martin. Monthly variation of forage and soil minerals in Central Florida. II. Trace Minerals. **Commun. Soil Sci. Plant Anal.,** 22: 1137-1149, 1991.

Ferguson, J.D. Nutrition and reproduction in dairy cows. In: The Veterinary Clinics of North America. **Fo. An. Pract. - Dairy Nutrition Management**, vol. 7, p. 483-507, 1991.

FERNANDES, M. H. M. R. **Body composition and nutritional requirements in protein and energy of goats with genetic makeup % Boer and ^ Saanen.** 2006. Thesis (Doctorate in Zootechnics)-Faculty of Agricultural and Veterinary Sciences, Unesp, 2006.

Filisetti, T.M.C.C.; Lobo, A.R. Dietary fiber and its effect on the bioavailability of minerals. In: COZZOLINO, S.M.F. **Biodisponibilidade de nutrientes**. 2° ed. Barueri: Manole. p.175-215, 2007.

Georgieviskii, V.I.; Annekov, B.N.; Samokhin, V.T. **Mineral nutrition of animals.** London, Butterworth, 1982.

GERASEEV, L.C.; et al.Body Composition and Nutritional Requirements in Calcium and Phosphorus for Gain and Maintenance of Santa Ines Lambs from 15 kg to 25 kg Live Weight. **Rev. bras. zootec.**, v.29, n.1, p.261-268, 2000.

Geraseev, l.c.; perez, j.r.o.; resende, k.t. Et al. Body composition and nutritional requirements for magnesium, potassium and sodium of santa ines lambs from 25 to 35 kg live weight. **Cienc. agrotec.**, Lavras, v.25, n.2, p.386-395, mar./abr., 2001.

Germano, R. H.; Barbosa, H.P.; Costa R.G. et al. Evaluation of the Chemical and Mineral Composition of Cactaceae in the Semi-arid Region of Paraiba. Agropecuária Técnica Vol. 20 N. 1, 1999.

Gomide, J. A. Mineral composition of tropical leguminous forage grasses. In: Simpósio Latino-Americano Sobre Pesquisa em Nutrigao Mineral de Ruminantes e Pastagens, 1, Belo Horizonte, 1976. *Proceedings...* Belo Horizonte: EPAMIG, p.20-33. 1976.

Grace, N. D. In: Managing trace element deficiencies: p. 9-32. **New Zealand Journal of Agricultural Research**, Palmerston North, New Zealand, 1994.

Grace, N.D., Ulyatt, M.J., Macrae, J.C. Quantitative digestion of fresh herbage by sheep. 3. Movement of Mg, Ca, P, K and Na in digestive tract. *J. Agric. Sci.,* v.82, p.321-330, 1974.

Graham, T. W. Trace element deficiencies in cattle. Vet. Clin. N. Am.: Food Anim. Pract. 7:153-215.,1991.

Gray, L. F., and L. J. Daniel. Some effects of excess molybdenum on the nutrition of the rat. **J. Nutr.** 53:43, 1954.

Grudtner VS, Weingrill P, Fernandes AL.. Aspects of absorption in the metabolism of calcium and vitamin D. **Rev Bras Reumatol** May/June; 37(3):143-151, 1997.

Guéguem L, Pointillart A. The bioavailability of dietary calcium. J Am Coll Nutrition. 19(2):119S-136S, 2000.

Guyton AC, Hall HE. Treatise on medical physiology. Rio de Janeiro: **Guanabara Koogan**; p.895-910, 1997.

Guyton AC. **Treatise on medical physiology**. 8.ed. Sao Paulo: Guanabara; 1991. p.316-8.

Haenlein, G.F.W. Cow and goat milk aren't the same--especially in somatic cell content. Dairy Goat J. 65 (12) : 806. 1987.

Haenlein, G.F.W., Goats: Are they physiologically different from other domestic food animals? Internat. Goat Sheep Res. 1: 173-175, 1980.

Haenlein, G.F.W., Role of goat meat and milk in human nutrition. Proceedings Vth International Conference on Goats, New Delhi, India, March 1-8, ICAR Publ., New Delhi, 2 (II): 575-580, 1992.

Hallberg L, Brune M, Rossander L. Iron absorption in man: ascorbic acid and dosedependent inhibition by phytate. **Am J Clin Nutr**. 49:140-4. 1989.

Hart, E. B., H. Steenbock, J. Waddell, and C. A. Elvehjem. Iron in nutrition. VII. Copper as a supplement to iron for hemoglobin building in the rat. **J. Biol. Chem.** 77:797, 1928.

Heaney RP, Abrams S, Dawson-Hughes B, Looker A, Marcus R, Matkovic V, et al. Peak Bone Mass. **Osteoporos Int**; 11:985-1009. 2000.

Heath, M. E.; Barnes, R. F.; Metcalfe, D. S. **Forage - The science of grassland agriculture**. Iowa, 643 p, 1985.

Hidiroglou, M., R. B. Carson, and G. A. Brossard. Influence of selenium on the selenium contents of hair and on the incidence of nutritional muscular disease in beef cattle. Can. J. Anim. Sci. 45:197, 1965.

Hungate, R.E. **The rumen and its microbes.** New York: Academic Press, p.346-347. 1966.

Hurley, W.L., Doane, R.M. Recent developments in the roles of vitamins and minerals in reproduction. **J. Dairy Sci.**, v.72, p.784, 1989.

Jones,T.C; Hunt,R.D.; King, N.W. **Pathología Veterinária**, 6ed., Sao Paulo, Manole Ltda, 2000, 1415 p.

Kabata-Pendias, A.; Pendias, H. **Trace elements in soils and plants**. Boca Raton: CRC, 315 p. 1984.

Kadu, M.; Kaikini, A. Prenatal development of caprine foetus. **Indian Journal of Animal Science**, v.57, n.9, p.962-9. 1987.

Kamada, H.; Hodate, K. Effect of dietary selenium supplementation on the plasma progesterone concentration in cows. **J. Vet. Med. Sci., Japan**. v. 60, n.1, p. 133-135, 1998.

Karn, James F., 2001. Phosphorus nutrition of grazing cattle: a review, **Animal Feed Science and Technology**, v.89, p.133-153, 2001.

Keener, H. A., G. P. Percival, and K. S. Marrow. Cobalt tolerance in young dairy cattle. J. Dairy Sci. 32:527, 1949.

Kessler, J. Mineral nutrition of goats. In: Morand-Fehr. (ed). Goat Nutrition. Pudoc Publi., Wageningen, Netherlands, EAAP Publ. No. 46, 104, 1991.

Khaled, N. F.; Illek, J. Influence of dietary supplementation of selenium-enriched yeast on selenium status of dairy goats. In: Congress On Macro End Trace Elements, 19th. 1999. Friedrich-Schiller University, Jena, Germany, Dec. 3-4, p. 244-251, 1999.

Khan, Z.I., A. Hussain, M. Ashraf, E.E. Valeem and I. Javed. Evaluation of variation of soil and forage minerals in pasture in a semiarid region of Pakistan. **Pakistan Journal of Botany**, 37: 921-931, 2005.

Khan, Z.I., M. Ashraf, I. Javed and S. Ermidou-Pollet. Transfer of sodium from soil and forage to sheep and goats grazing in a semiarid region of Pakistan. Influence of the seasons. **Trace Elements Electrolytes**, 24: 49-54, 2007.

Khan, z.i.; ashraf, m.; ahmad, k. Et al. Evaluation of mineral composition of forages For grazing ruminants in pakistan. **Pak. J. Bot.**, 41(5): 2465-2476, 2009.

Kozloski, G. B. **Biochemistry of ruminants**. Santa Maria: Federal University of Santa Maria, 2002. 139p.

Krause MV, Mahan LK. Nutritional care in anemia. In: Krause MV, Mahan LK. **Food nutrition and diet therapy**. 7.ed.Sao Paulo: Roca, p.581-8, 1991.

Larvor, P. 28Mg kinetics in ewes fed normal or tetany prone grass. Cornell Vet. 66:413, 1976.

Leng, R.A. 1990. Factors affecting the utilization of "poorquality" forages by ruminants particularly under tropical conditions. **Nut. Res. Rev.**, 3(3):277-303.

Linder, M. C. Nutrition and Metabolism of the trace elements. In: Nutritional Biochemistry and Metabolism with clinical applications. Second edition. **Elsevier Science Publishing Company**. New York, New York C.7, p. 215-276, 1991.

Lofgreen, G. P.; Kleiber, M. The availability of the phosphorus in alfalfa hay. Journal of Animal Science, Champaign, v. 12, n. 2, p. 366-371, 1953.

Lofgreen, G.P.; Garrett, W.N. **A System for Expressing Net Energy Requirements and Feed Values for Growing and Finishing Beef Cattle.** J ANIM SCI 27:793-806. 1968.

Luberda, Z. The role of glutathione in mammalian gametes. **Reprod. Biol.**, Olsztyn, Poland, v. 5, n. 1, p. 5 - 17, 2005.

Lynch, S. R. Interaction of iron with other nutrients. **Nutrition Review,** Lawrence, v. 55, n. 4, p. 102-110, apr. 1997.

Macdonald, R. S., "The role of Zinc in growth and cell proliferation", J. Nutr., 130 (May), (5S Suppl):1500S-8S, 2000.

Maynard. D.G., Stewart. J.W.B. and Bettany, J.R., Theeffects of plants on soil sulfur transformations. **Soil Biol. Biochem.**, 17: p.127-134 1985.

Mcdowell, L. R. Feeding minerals to cattle to pasture. **Animal Feed Science and Technology**, v. 60, n. 3-4, p. 247-271, 1996.

McDowell, L.R and Arthington, J.D. Minerals for Grazing Ruminants in Tropical Regions, 5th Ed. 86pp. University of Florida, Gainesville, Fl, 2005.

Mcdowell, L.R. **Minerals for grazing ruminants in tropical regions, emphasizing Brazil.** 3 ed., University of Florida , 92 p., 1999.

Mcdowell, L.R. **Minerals in animal and human nutrition**. Gainesville: Academic Press, 524 p. 1992.

Mcdowell, L.R.; Conrad, J.H.; Ellis, G.L.; Loosli, J.K. **Minerals for grazing ruminants in tropical regions.** Gainesville: University of Florida, 86p, 1983.

Mcdowell, L.R.**Vitamins in animal and human nutrition**. Iowa: Iowa State University Press, 793p. 2000.

Mcdowell, L. R. **Nutrition of grazing ruminants in warm climates**. San Diego : Academic, 443 p. 1985.

McHargue, J. S. The association of copper with substances containing the fatsoluble A vitamin. Am. J. Physiol. 72:583, 1925.

Mestchy, F. Recent progress in the assesment of mineral requirement of goats. **Liv. Prod. Sci.**, Paris, France, v. 64, p. 9-14, 2000.

Michell, A. R. The clinical biology of Na: the physiology and pathophysiology of Na$^+$ in mammals. **Pergamon Press**, Oxford, United Kingdom. 1995.

Miller, J.K., Ramsey, N., Madsen, F.C. The trace elements. In D.C. Church. **The Ruminant Animal. Digestive Physiology and Nutrition**. Waveland Press, Inc. pp. 342400. 1993b.

Miller, J. I., and K. Tillapaugh. Iodide Medicated Salt for Beef Cattle. Cornell Feed Service No. 62. Cooperative Extension Service, Cornell University, Ithaca, N.Y, 1966.

Mills, C.F. Biochemical and physiological indicators of mineral status in animals: copper, cobalt and zinc. **Journal of Animal Science**, v.65, p.1702-1711, 1987.

Minson, D.J. **Forage in ruminant nutrition**. New York: Academic Press, 1990. 483 p.

Morse, D.; Head, H. H.; Wilcox, C. J. Disappearance of phosphorus in phytate from concentrates in vitro and from rations fed to lactating dairy cows. **Journal of Dairy Science**, Champaign, v. 75, n. 7, p. 1979-1986, 1992.

Morais, S.S. **Importance of Mineral Supplementation for Beef Cattle**. Embrapa Beef Cattle. Document 114, Campo Grande, 2001.

National Research Council NRC. Subcommittee on Selenium. **Selenium nutrition.** Washington, DC.: National Academic Press. 1983.

National Research Council - NRC. **Nutrient requirements of beef cattle.** 7.ed. Washington, DC: National Academy Press. 242p. 1996.

National Research Council - NRC. **Nutrient requirements of beef cattle**. 7. ed REVISED. Washington, DC: National Academy Press, 232 p. 2000.

National Research Council - NRC. **Nutrient requirements of sheep.** 5.ed. Washington, D.C.: 1975. 69p.

National Research Council - NRC. **Nutrient requirements of small ruminants: Sheep, goats, deer and new world camelids.** Washington, DC: National Academy Press.362p, 2007.

NATIONAL RESEARCH COUNCIL - NRC. **Nutrient requirements of the dairy cattle.** 6.ed. Washington, D.C., 158p. 1989.

National Research Council NRC. **Mineral Tolerance of Domestic Animals**. Natl. Acad. Sci., Washington, D.C. 1980.

National Research Council. **Energy for rural development renewable resourses and alternative technologies fos developing countries.** Washington: Academy Press, 238p. 1981.

National Research Council. **Nutrient requirements of beef cattle**. Minerals, 6. ed. rev. Washington: National Academic Press, 1984. p. 11-23.

National Research Council - NRC. **Nutrient requirements of sheep.** 6.ed. Washington, DC: National Academy Press 99p, 1985.

Nóbrega, G.H.; Silva, A.M.A.; Pereira Filho, J.M. et al. Body composition and macromineral requirements for weight gain in grazing goats **Acta Sci. Anim. Sci.** Maringá, v. 31, n. 1, p. 69-75, 2009.

Oblitas, F.et al. Effect of selenium supplementation on blood glutathione peroxidase (GSH-Px) activity and weight gain in selenium-deficient cattle kept at pasture. **Arch. Med. Vet.,** Valdivia, Peru, v. 32, n. 1, p. 1-11, 2000.

Ogebe, P.O. and J.A. Ayoade, L.R. McDowell, F.G. Martin and N.S. Wilkinson. 1995. Mineral concentrations of forages and soils in Benue State, Nigeria. I. Macrominerals and forage *IN VITRO* organic matter digestibility and crude protein concentrations. Communications in. Soil Science and Plant Analysis, 26: 1989-2007.

Paulino, M.P.; et al. Body composition and requirements for mineral macroelements (Ca, P, Mg, Na and K) of cattle from four zebu breeds. **Rev. bras. zootec.,** v.28, n.3, p.634641, 1999.

Pedreira, M.S.; Berchielli, T.T. Minerals. IN: Berchielli, T.T., Pires, A.V., Oliveira, S.G. Nutrigao de Ruminantes. Jaboticabal: FUNEP, 583P, 2006.

Peixoto, A.M.; Moura, J.C.; Faria, V.P. Anais do 3° Simpósio sobre Nutrigao de Bovinos. FEALQ. **Proceedings...** 145p. 1985.

Peixoto, A.M.; Moura, J.C.; Faria, V.P. de. **Cattle nutrition - Basic and applied concepts.** 7v. Ed.5^a , Piracicaba: FEALQ, p. 414-415, 1995.

Pineda O. El uso del hierro aminoquelado en el control de La deficiencia de hierro y de anemia ferropriva. In: **Resúmenes de lo congreso latinoamericano de nutrición**; Nov

15; Caracas. 1994.

Postgate, J. R. The sulphur cycle, p. 259. *In* G. Nickless (ed.). Inorganic Sulphur Chemistry. Elsevier Publishing Co., New York, 1968.

Pott, E.B. **Mineral and crude protein contents in forage plants from the uplands of Corumbá, MS**. Technical Com. N.006/86, 1986.

Powell, K.; Reid, R. L.; Balasko, J. A. Performance of lambs on perennial ryegrass, smooth bromegrass, orchardgrass and tall fescue pastures. II. Mineral utilization, *IN VITRO* digestibility and chemical composition of herbage. **Journal of Animal Science**, Champaign, v. 46, n. 6, p. 1503-1514, 1978.

Prabowo, A., L.R. McDowell, N.S. Wilkinson, C.J. Wilcox and J.H. Cornad. Mineral status of grazing cattle in South Sulawesi, Indonesia; I. Macrominerals. *ASIAN-AUST. J. NIM. SCI.,* 4: 111-120, 1990.

Queiroz, A. C.; Gouveia, L. J.; Perreira, J. C. et al. Nutritional Requirements of Growing Alpine Goats. 1. Nutritional Requirement of Phosphorus for Maintenance: Endogenous Losses and Comparative Slaughter. **Revista Brasileira de Zootecnia**, v. 29, n.4, p.12051215, 2000.

Rayssiguier, Y., J. M. Garel, M. J. Prat, and J. P. Barlet. Plasma parathyroid hormone and calcitonin levels in hypocalcaemic magnesium deficient calves. Ann. Rech. Vet. 8:267, 1977.

Resende, K. T.; Ribeiro, S. D. A.; Dorigan, C. J.; Carvalho, F. F. R.; Costa, R. G.; Vasconcelos, V. R. Goat nutrition: new systems and nutritional requirements. In: Lima, F. A. M.; Leite, E. R.; Martins Filho, R.; Villarroel, A. B. S.; Fernandes, A. A. O.; Oliveira, S. M. P.; Moura, A. A.. (**Annual Meeting of the Brazilian Society of Animal Science**. Fortaleza: SBZ, v. 1, p. 77-99, 1996.

Resende, K.T. **Methods for estimating body composition and nutritional requirements for protein, energy and inorganic macroelements in growing goats.** Vigosa: UFV, 1989. 130p. Thesis (Doctorate in Zootechnics) - Federal University of Vigosa, 1989.

Ribeiro, S.D.A. Caprinocultura: **Rational Breeding of Goats**. Sao Paulo. Nobel, 1997. 1ª Ed. 318 p, 1997.

Ribeiro, S.D.A. **Body composition and protein, energy and macromineral requirements of crossbred goats in the initial growth phase.** Jaboticabal: FCAVJ/UNESP, 1995. 100p. Dissertation (Master's Degree in Zootechnics) - Universidade Estadual Paulista, 1995.

Richert, D. A., and W. W. Westerfield. Isolation and identification of the xanthine oxidase factor as molybdenum. **J. Biol. Chem.** 203:915, 1953.

Riordan, J. F., and B. L. Vallee. Structure and function of zinc metalloenzymes, pp. 227256, *In* A. S. Prasad (ed.). Trace-Elements in Human Health and Disease, vol. I. Academic Press, New York, 1976.

Rook, J. A. F. and J. E. Storry. Magnesium in the nutrition of farm animals. **Nutr. Abstr. Rev.** 32 : 1055, 1962.

Rosa, I. V. **Microelement deficiencies and reproduction**. Campo Grande: Embrapa Gado Corte. 9 p. 1993.

Rosado, M. **Effect of the fatty acid-calcium complex on apparent digestibility, some rumen parameters and passage rate in lactating cows.** Vigosa: Federal University of Vigosa, 1991.s 96p. Dissertation (Master's Degree in Zootechnics) - Federal University of Vigosa, 1991.

Rosol,C.; Capen, C. Calcium-regulating hormones and diseases of abnormal mineral (calcium, phosphorus, magnesium) metabolism. In: Kaneko J.J (Ed.) **Clinical Biochemistry of Domestic Animals.** 5.ed. New York: Academic Press, 1997.

Runho, R. C.; Gomes, P. C.; Rostagno, H. S. et al. Available phosphorus requirements for male and female broilers from 1 to 21 days of age. **Revista Brasileira de Zootecnia**,

Vinosa, v. 30, p. 187-196, 2001.

Sanderberg A, Anderson H, Carlson NG, Sandstrom B. Degradation products of bran phytate formed during digestion in the human small intestine: effect of extrusion cooking on digestibility. **J Nutr**; 117:2061, 1987.

Santos, Y. C. C. **Body composition and nutritional requirements of energy and protein of Bergamácia lambs from 35 to 45 kg live weight**. Thesis (Master's Degree in Zootechnics) - Federal University of Lavras, Lavras. 63 p, 2000.

Silva, F.F.; Valadares Filho, S.C.; Ítavo, L.C.V.; et al. Net and dietary requirements of energy, protein and mineral macroelements of beef cattle in Brazil. **Revista Brasileira de Zootecnia**, v.31, n.2, p.776-792, 2002.

Silva, J.F.C. Inorganic macro-nutrient requirement in cattle: the ARC/AFRC system and the Brazilian experience. In: INTERNATIONAL SYMPOSIUM ON RUMINANT NUTRITIONAL REQUIREMENTS, Vigosa, 1995. **Proceedings...** Vinosa, p.311-343. 1995.

Smith, G.S.; Middleton, K.R.; Edmonds, A.S. Sodium Nutrition of Pasture Plants. **The New Phitologist**. 1980.

Smith, O. B.; Akinbamijo, O. Micronutrients and reproduction in farm animals. **Anim. Reprod. Sci.** Ottawa, Canada. v. 60/61, p. 549-560, 2000.

Solaiman, S.G.; Maloney, M.A.; Qureshi, M.A. et al. Effects of high copper supplements on performance, health, plasma copper and in goats. **Small Ruminant Research** 41. 127139, 2001.

Sousa J.C., Gongalves E.M., Viana J.A.C. & Darsie G. Mineral deficiencies in cattle from Roraima, Brazil. IV. Magnesium, sodium and potassium. **Pesq. Agropec. Bras.** 22(1):89. 98, 1987.

Souza, H.M.H, Queiroz, A.C., Resende, K.T. et al. Nutritional requirements of growing Alpine goats. 1. nutritional requirement of calcium for maintenance. **R.Bras.Zootec.**, v.27, n.1, p. , 1998a.

Souza, H.M.H, Queiroz, A.C., Resende, K.T. et al. Nutritional requirements of growing alpine goats. 2. Body composition and weight gain in protein, ether extract, energy, calcium and phosphorus. **R.Bras.Zootec.**, v.27, n.1, p. , 1998b.

Sparks, D.L. Chemistry of soil potassium in Atlantic coastal plain soils: a review. **Communications in Soil Science and Plant Analysis**, New York, v.11, p.435- 449, 1980.

Spears, J.W. Minerals in forage. In: Fahey J.R., G.C. (Ed.) Forage quality, evaluation and utilization. **American Society of Agronomy**. National Conference On Forage Quality, Evaluation And Utilization, p.281-317, 1994.

SUTTLE, N.F. Copper deficiency in ruminants; recent developments. **Vet. Rec.**, v.119, p.519-522, 1986.

Suttle, N.F. The interactions between copper, molybdenum, and sulphur in ruminant nutrition. **Annual Veterinary Nutrition,** v.11, p.121-140, 1991.

Suttle, N.F. The interactions between copper, molybdenum, and sulphur in ruminant nutrition. **Annual Veterinary Nutrition,** v.11, p.121-140, 1991.

Teixeira, J.C. Nutrigao de Ruminantes. Edigoes FAEPE, Lavras, MG, 239p. 1992.

Tejada, R., L.R. McDowell, F.G. Martin and L.H. Conrad. Evaluation of the macromineral and crude protein status of cattle in specific regions in Guatemala. **Nutr. Rep. Int.**, 35: 989- 998, 1987.

Ternouth, J.H.; Sevilla, C.L. Dietary calcium and phosphorus repletion in lambs. **Australian Journal of Agricultural Research** v.41, n.2, p.413-420, 1990.

Thomas, J. W. Mechanisms responsible for grass tetany, p. 14. *In* Proc. Ga. Nutr. Conf. Feed Manuf. 1965.

Thompson, D.J. **Potassium in animal nutrition**. Libertyville, International Minerals and Chemical Corporation, 1972.

Tiffany, M.E., L.R. McDowell, G.A. O'Connor, F.G. Martin, N.S. Wilkinson, E.C. Cardoso, S.S. Percival and P.A. Rabiansky. Effects of pasture applied bio solids on performance and mineral status of grazing beef heifers. *J. Anim. Sci.,* 78: 1331, 2000.

Tiffany, M.E., L.R. McDowell, G.A. O'Connor, H. Nguyen, F.G. Martin, N.S. Wilkinson and N.A. Katzowitz. Effects of residual and reapplied biosolids on forage and soil concentrations over a grazing season in north Florida. II. Microminerals. Communication in Soil. **Science and Plant Analysis***, 32: 2211-2226, 2001.

Tiffin, L. O. Translocation of micronutrients in plants. In: Mortvedt, J. J.; Giordano, P. M.; Lindsay, W. L. (Ed.). **Micronutrients in agriculture**. Madison: Soil Science Society of America, p. 199-229, 1972.

Tinker, P. B. Levels, distribution and chemical forms of trace elements in food plants. **Philosophical Transactions of the Royal Society of London, Series B, Biological Sciences**, v. 294, n. 1071, p. 41-55, 1981.

Tisdale, S.L.; Nelson, W.L.; Beaton, J.D.; Havlin, J.L. **Soil fertility and fertilizers**. 5.ed. New York : MacMillan, 1993. 635p.

Tokarnia, C.H., Dobereiner, J. Experimental poisoning in sheep by myomium (*BACCHARIS CORIDIFOLIA)*. **Pesquisa Veterinária Brasileira**. 11:19-6,1976.

Tomas, F. M. & Potter, B. J. T. he site of magnesium absorption from the ruminant stomach. **British Journal of Nutrition** 36, 37-45, 1976.

Tucker, H. F., and W. D. Salmon. 1955. Parakeratosis in zinc deficiency disease in pigs. Proc. Soc. Exp. Biol. Med. 88:613.

Underwood E.J. & Suttle N.F. **The Mineral Nutrition of Livestock**. 3 ed. CABI Publ. Wallingford.1999, 614 p.

Underwood, E. J. Los minerales en la nutrición del ganado. Zaragoza, 209 p, 1983.

Underwood, E. J.; Suttle, N. F. **The mineral nutrition of livestock**. 3.ed. Wallingford: Cabi Publishing, 1999. 614 p.

Underwood, E.J. **The mineral nutrition of livestock**. London: Academic Press, 1981.

Underwood, E.J. **Trace elements in human and animal nutrition**. 3.ed. New York: Academic Press, 1971.

Underwood, E.J. **Trace elements in human and animal nutrition**. 4.ed. London: Academic Press, 545p, 1977.

Valadares Filho, S.C. et al. Total and partial apparent absorption of sodium, potassium, magnesium, copper and manganese in cattle fed purified and semi-purified feed. In: Reuniao Anual Da Sociedade Brasileira De Zootecnia, 28, Joao Pessoa, 1991, **Anais...**Joao Pessoa:SBZ, p.179 1991.

Vallee, B. L. The entatic properties of cobalt carboxypeptidase and cobalt procarboxypeptidase. *In* W. G. Hoekstra, J. W. Suttie, H. E. Ganther, and W. Mertz, eds. Trace Element Metabolism in Animals-2, p. 5. **University Park Press, Baltimore**, Md. 1974.

Van Soest, P.J. **Nutritional ecology of the ruminant**. Ithaca: Cornell University Press, 1994, 476p.

VanNiekerk, F. E; Cloete, SWP; Heine, EWP; vanderMerwe, A; duPlessis, SS; BEKKE, D. The effect of selenium supplementation during the early post-mating period on embryonic survival in sheep. **J. South African Vet. Assoc.**, Stellenbosch, South Africa, v. 67, n. 4, p. 209-213, 1996.

VÉRAS, A.S.C. **Minerals in forage.** Seminar (ZOO 471): Ruminant Nutrition II: UFV. 1999. 21p.

Watson, L. T., C. B. Ammerman, J. P. Feaster and C. E. Roessler. Influence of manganese intake on metabolism of manganese and other minerals in sheep. **J. Anim. Sci.** 36:131, 1973.

Whitehead, D. C.; Goulden, K. M.; Hartley, R. D. The distribution of nutrient elements in cell wall and other fractions of herbage of some grasses and legumes. **Journal of the**

Science of Food and Agriculture, London, v. 36, p. 311-318, 1985.

Wilson, J.G. Bovine functional infertility in Devon and Cornwall: response to manganese therapy. **Vet. Rec.,** v. 9, p.562-566, 1966.

Witt, K.E.; Owens, F.M. Phosphorus ruminal availability and effects on digestion. **Journal of Animal Science**, v.56, n.4, p.930-937, 1983.

Young, L., and G. A. Maw. The Metabolism of Sulphur Compounds. John Wiley & Sons, New York, 1958.

Zervas, G., Nikolaoy, E. and Manzios, A. Comparative study of chronic copper poisoning in lambs and young goats. Proc. 6th Intern. Trace Element Symposium, University Leipzig-Jena, Germany, 569.,1989.

Zhou JR, Erdman, FWJ. Phytic acid in health and disease. Critical reviews in food science and nutrition; 35(6):495-508, 1995.

Printed by Books on Demand GmbH, Norderstedt / Germany